On Integration in Plants

Rudolf Dostál

On Integration in Plants

BY

RUDOLF DOSTÁL

Translated by Jana Moravkova Kiely

Edited by Kenneth V. Thimann

Harvard University Press, Cambridge, Massachusetts

1967

Distributed in Great Britain by Oxford University Press, London

Library of Congress Catalog Card Number 67-27083

Printed in the United States of America

Translated from *O Celistvosti Rostliny*
(Prague: State Agricultural Publishing House, 1959)

For J.M.

Preface

I was led to writing this book, which is based mainly on my own experiments on integration in plants, by a number of circumstances. Both in popular lectures and in scientific meetings my work has aroused much interest and I have been asked many times to publish all the results together. In an article published in the journal *Science and Life*, entitled "Is there also a regulatory mechanism in plants?" I briefly outlined my conception of the main factors of plant integration. However, when in October 1956 I received from Professor M. H. Chailakhian, who is well known for his interesting experiments on photoperiodism, the book *Integrity of the Organism in the Plant World*,[1] in which for the first time he analyzes this most important of biological problems, I decided to write this book. Just as a doctor has to consider the human organism as a whole, so that treatment of a single part will heal and not harm the whole, so also the grower has to consider the plant organism in its wholeness, since any interference, be it a change in nutrition or an operation, has repercussions throughout the whole plant.

In my work I have tried, more perhaps than did Professor Chailakhian, to comprehend the interactions among the dif-

[1] *Celostnost organizma v rastitelnom mire*, 1955.

ferent parts of the plant; that is to say, the correlations through which the various parts of the plant body are linked together into a harmonic whole, even though they lack the nervous and hormonal system possessed by animals. Most such basic problems of the living organism are more easily studied in plants than in animals, and it is not surprising that biologists turn to plants so often, even though previously they may have mostly studied animals. In his last experiments, Charles Darwin and his son Francis supplied the basic idea for the discovery of growth "regulators," which are now produced in great quantities in chemical industries for the benefit of agriculture. Similarly, the American animal physiologist, Jacques Loeb, ended his scientific work by extensive experiments on the tropical plants of the Bryophyllum family, in an attempt to comprehend the dynamics of regeneration processes in this uniquely suitable experimental plant so admired by Goethe.

Here again, no real understanding is possible without experimentation. It is not enough to observe the usual state of plants in nature or in culture and to speculate about it. "The knowledge of causes is the only true one," said Francis Bacon, the victor over medieval scholasticism. It seems, however, that the more I have experimented with living plants, the farther away I have moved from my goal of understanding the hidden causes; at each step new elements arise, which have been hidden in every particle of the living matter of plants during their phylogeny and which, up to now, neither biophysics nor biochemistry has been able to reveal. And yet such knowledge would have the most fundamental meaning. Of course, discoveries of details of the laws of nature can be extremely useful in practice. As Chaplygin, hero of socialist work, says, "Scientific work is not a dead scheme, but a ray of light for practice." The reason why I have not always used agricultural plants for my experiments is that they are not always as suitable for experimentation as certain undomesticated plant types. The most interesting results have been those obtained from the acellular Mediterranean alga *Caulerpa prolifera*. In the large undivided mass of living matter of this alga the same laws of integration are valid as in the highly organized trees, which can reach their large size and longevity only with the help of a complex cellular structure.

My first experiment with plants was performed successfully more than fifty years ago in the Institute for Plant Anatomy and Physiology of that unsurpassed teacher, Academician Bohumil Němec.[2] In this I found that pea seedlings, from which one cotyledon and the main stem had been removed, always replaced the stem by a shoot on the side of the removed cotyledon. From this I acquired the certitude that all regulations in plants take place with an absolute determinism, an iron necessity, and that it is up to us gradually to discover all these laws and not to let ourselves be swayed by the empty explanations of idealists. Only a dialectico-materialistic philosophy and method can lead us to this goal, and its ultimate attainment will be of primary benefit to agricultural practice.

*   *   *   *   *

From a thorough and unbiased study of living organisms modern biology has drawn two basic laws: the law of the unity between ontogeny and phylogeny, and the law of unity between the organism and its environment. Without the first law we cannot understand satisfactorily the regularity of the development of the living organism. *Ontogeny*, or the development of each individual organism, and *phylogeny*, or the development of the species from its oldest acellular ancestors (or rather from unorganized living matter) through a succession of transitory ontogenies up to the forms of today, really

[2] At the age of 90 Professor Němec was still carrying on experiments, and indeed he attended the International Botanical Congress (Edinburgh, 1964). Subsequently he chaired the Organizing Committee of the Mendel Memorial Symposium, Brno, August 1965. (Ed.)

constitute two different views of the same evolution of living organisms. This basic conception of the continuity of life from its beginning on earth to the present time springs from the immortal work of Charles Darwin (1859), and has become one of the two pillars of Michurinian biology. The other pillar, as noted above, is the law of the unity between the organism and its environment, which is now and has always been the main factor in the shaping of organisms.

Both of these laws have been beautifully substantiated by the sixty years of successful experimentation, especially in fruit growing, of I. V. Michurin (1948), who therefore can truly be called the great re-creator of nature. Michurin showed clearly when and how one must interfere with the development of the organism in order to ensure that the changes brought about by experiment become hereditary. By his long and indefatigable experimental work he laid the foundation of creative Soviet Darwinism.

Without these two basic laws it would be impossible to understand the integration of an organism as a whole — a major biological problem both from the theoretical and from the practical point of view.

Living beings, especially those with a complex organization, are composed of organs, tissues, and cells. However, they are not a mere summation, an aggregate of all these components, for throughout their development they appear as a harmonious whole. It is only in this wholeness that the various components find favorable conditions for their development and function, and only so can they fully contribute to the preservation of the individual and of the race.

The integration of the organism was given a sound materialistic basis by Charles Darwin, when in his classical work, *Origin of Species by Means of Natural Selection*, he wrote that all parts of the organism are linked together by more or less tight correlations which arise during their individual development, and which after many generations are transformed into hereditary characteristics. He thus stressed the historical origin of correlations and designated them as being based on heredity. Any metaphysical conception of the integration of an organism, to which one might be tempted by a superficial observation of nature, is ruled out.

Once, in August 1838, after lunch, two German naturalists, a botanist, Franz Schleiden, and a zoologist, Theodor Schwann, were looking in the laboratory at microscopic preparations of plant and animal tissues. They were amazed at a certain similarity between these tissues, namely that both were composed of cells, and they brought forth the theory that cells are the elementary building units of both types of organism. In spite of its erroneous nature this theory gave new incentive to further studies of the anatomy of plants and animals. It was stressed by the German pathologist Virchow (1871), who called the cell the ultimate morphological unit of all life, arising only from pre-existing cells through division. The organism is, according to him, a cellular State in which the components, that is to say the cells, have more importance than the organism as a whole. This is a basically doubtful conception, for in reality one has to view the parts and the whole as the union of two opposites. The error of the cell theory, thus understood, is obvious from the fact that even in acellular, sometimes highly organized plants, such as the marine alga *Caulerpa prolifera* (Fig. 7a), we find biological mechanisms of integration very similar to those in cellular plants.

Virchow's conception was further developed by the German physiologist Verworn (1901), according to whom a large organism can never be made of just one cell, but must be composed of a large number of cells, a whole State of cells, in which each cell must sacrifice some of its independence for the common good. This, according to him, brings about an even greater variety between cells than between individuals of human society, and human society could apparently draw many good lessons for its social reforms from the organization of this cell State. Expanding his speculations even further, Verworn compares animals, with their highly developed central nervous system, to a despotic monarchy, while plants, with no nervous system, are compared to a popular republic. Just as Virchow is considered the founder of cell pathology, so Verworn has been called the founder of cell physiology. His conception of the living body as a federation of cells, tissues, and organs was also shared by Danilevskii (1913).

These erroneous conceptions were ruled out by F. Engels (1925) in the *Dialectic of Nature*, in which he showed that neither the mechanical union of bones, blood, cartilages, and various tissues, nor the chemical union of various compounds, could constitute a living body. The organism actually appears as if it did not have components.

The great physiologist I. P. Pavlov (1952) in his study of conditioned and unconditioned reflexes, showed that in higher animals all functions are connected with the activity of the brain cortex, and this ensures the integration of the organism. Bykov (1950) evaluated this concept as follows: — "I suspect that the transition from the understanding of an organism as a summation of separately studied organs and systems, to the pavlovian conception of an integrated whole in its relations to the outside environment, was a triumph of scientific thinking equal to that represented by the shift from the ptolemaic conception of the universe (according to which the earth is in the middle of the universe and all the planets, sun included, revolve around it) to the theory of the solar system of Copernicus."

The organism as a whole cannot be separated from its environment; it is only through the union of both that the normal development of the organism is ensured. The founder of Russian physiology, Sechenov (1952), expressed this by saying that everywhere and always life is composed of two factors: a certain organization which can be varied, and the external factors which influence it. The organism does not possess within itself the conditions for its existence. It is constantly under the influence of its environment, without which it cannot exist. The question as to which is more important for life, the environment or the body proper, is meaningless. The organism lives only so long as everything bears the right relationship, not only within it, but also between it and the environment (Chailakhian, 1958).

It is our task to continue through experiment to bring solutions to the general problem of integration. As in many other biological problems, plants appear to be more suitable tools than animals for this study. This is partly due to the fact that even in the most highly organized plants all the organs can be reached from the outside, and it is not necessary to cut deeply into their bodies to be able to follow the functioning of their organs.

Typical green plants are specialized in nature for a diet of mineral substances, which they take up from the soil through the roots and from the air through the leaves, and then transform into organic substances such as sugars, starch, fats, proteins, and many others. They thus differ radically from animals, which need the organic substances of plants for their nutrition. On the other hand the differentiation of the animal body is more complicated; in plants we would look in vain for governing centers comparable to the brain, controlling the development and functioning of organs and of their components, the tissues and cells. This integration is ensured by the nervous and neurohormonal system, while plants have no real nerves or nerve-hormones comparable to those in animals.

In plants the problem of integration might even appear pointless, since even the most highly organized plants, such as the trees, form a great number of apparently similar parts; these can be easily reduced to a smaller number without apparent harm, by the cutting off of some of them, as is commonly done in tree nurseries. At the other extreme, an isolated part can often grow into a whole plant, a behavior which is commonly made use of in vegetative multiplication, which is for many purposes preferable to sexual reproduction by seed.

In plants, therefore, divisibility prevails over integration, since isolated individual parts can under favorable conditions replace what they lack to re-form a whole. This power protects plants against unfavorable conditions, to which they are much more exposed than animals, since plants are fixed throughout their lives to a definite location.

Experience shows that this power to divide and reproduce vegetatively is not the same in all plants. We find it, however, even in the most phylogenetically advanced plants, while this power is lacking in higher animals, such as man and other mammals. Plants form a large number of vegetative primordia, for example, adventitious buds, against the eventuality of damage to the aerial system. For this reason the problem of integration has a quite different meaning in plants and in animals. In plants the problem is to explain why all of these countless vegetative primordia do not develop at the same time, why only some of them grow and the others remain inhibited. Such behavior has tremendous importance for the plant, for it ensures a harmonious development, with the formation in any given condition of a sufficient number both of primordia and of reserves for the next generation. Herbaceous plants first develop roots and leaves and only then produce fruits, seeds, tubers, bulbs, rootstocks, and other reserve organs, while in woody plants growth stops after a short time and the products of assimilation are stored in the roots, trunks, and branches for the benefit of the next vegetative season. All these phenomena depend on inhibitions, which exert the main role in plant correlations and which Darwin recognized as the basis of integration. We shall therefore follow the development of correlations in plants from the very beginning of development, the embryonic phase.

Rudolf Dostál

Contents

Illustrations

Foreword

Rudolf Dostál, who celebrated his eightieth birthday in 1965, has been in many ways a pioneer, but one whose work has been little recognized in western countries. Ever since his first paper in 1908, "Korelační vztahy u klíčních rostlin Papilionaceí" (Correlative relations in leguminous plants), his major interest has been in the influence of one plant organ upon the others, for example, the relations between leaves and buds, between leaves and tubers, between the two cotyledons. or between root and shoot. His important but not well-known experiments on the growth-regulating action of leaves, published in 1926, and an earlier study of 1909, have influenced a small number of workers, myself included, to think about the general problems of interrelation and control within the plant, and thus have fed directly into the field of hormone action. Robin Snow's experiments at Oxford on the control exerted by the growing apex over the lateral buds (Apical Dominance) were strongly influenced by Dostál, and they led rapidly to the recognition of auxin as a prime mover in this dominance. Dostál's continued interest in the subject and his ever broadening concern with other examples of "correlation," such as the shedding of needles in evergreens, branching patterns in trees and shrubs, and the morphogenesis of the marine alga Caulerpa, are demonstrated by his 111 papers appearing in an almost unbroken series from 1908 to 1966. (A complete bibliography of these recently appeared in *Biologia Plantarum*, Prague, *6*, 242, 1965.)

It was of particular interest, therefore, when I decided in 1961 (in the course of a journey to the Soviet Union) to pay a visit to Professor Dostál, to find that he had recently written a small book, *O Celistvosti Rostliny*, embodying not only his published observations but also a great quantity of original and as yet unknown work. In this his experiments on a variety of specific and local influences have culminated in a single concept: The Integration of Plants. It should be added that Professor Dostál has not attended international congresses very often; like his more famous predecessor and fellow townsman, the abbot Gregor Mendel (1822–1884), he has spent his life in Brno. Many of his most important contributions have been presented to the Moravian Academy of Sciences, in Brno. Here they have attracted no more immediate attention than did those of Mendel, which appeared in the proceedings of the Brno Society for Nature Research. It was natural, therefore, that I should suggest to him the desirability of bringing this material to the attention of the general fraternity of experimental botanists, by way of an English translation. Dr. Dostál agreed, and some time after my return to Harvard I was so fortunate as to be able to interest Jana Kiely (then Miss Moravkova) in undertaking the literal translation, which I then edited for English idiom and scientific content. The work proceeded slowly, since both of us were busy, although it was never anticipated that it would spread over as much as four years. During this time a few of the outdated sections have been modified by Dr. Dostál, and his changes have been worked into the text throughout. My editorial comments have been restricted to footnotes.

The book should be of interest to western readers for quite

another reason. Russian biology has until very recently, and for over thirty years, been dominated by the non-Mendelian genetics of Michurin, Lysenko, and their followers. Their views on the mechanism of inheritance are not particularly pertinent to Dostál's work, but the way in which they envisage stock-scion relationships is extremely pertinent. This book gives us an opportunity to see how greatly the views held in the Soviet Union have influenced those held in the "satellite" countries. The perceptive reader will note how often, even though Lysenko and Michurin may not be directly referred to, the argument is tinged with, or even modified by, the Lysenko position. Thus it provides an invaluable outlook on to an area of biological history which some day will attract concentrated attention — the area of administrational control over scientific views.

One can only conclude with the conventional hope that the reader with horticultural or botanical interest, whether lay or professional, may find in this book some stimulus to fresh thought about plants. Perhaps the student or professional researcher may find that it leads him directly to new experiments, for there are literally dozens of lines of investigation laid open here to the developmental biologist. The recognition that plant growth and development are under the control, not of auxin alone, but of a number of quite different types of hormones, has opened up a field which I have referred to elsewhere as the Endocrinology of Plants, and another way of viewing the same phenomenon is what Dostál calls the Integration of Plants. It is certainly a field of great importance and promise.

Kenneth V. Thimann

On Integration in Plants

CHAPTER I

Integration at the Embryonal Level

Plants differ widely in external shape and internal anatomical structure, although the major functions such as photosynthesis and respiration are basically the same in all of them. The process of photosynthesis is identical in the pine needle and in the oak leaf, although these differ a great deal in shape and structure.

The origins of these morphological differences are contained directly in the living matter which J. E. Purkyně called protoplasm. This is the living fluid content of cells, in which one usually distinguishes the cytoplasm, the nucleus, plastids, and mitochondria. The result of their cooperation is life itself, which is always linked to some degree of differentiation of the components of the living body and to a certain shape and structure. In each cell — to a certain extent even in each little cytoplasmic particle — the power to create the whole plant is enclosed. This statement is especially valid for embryonic cells, completely filled with protoplasm, without waste material or thick membranes. Such cells form the tissues of the tips of stems and roots, called apical meristems. In higher plants, also, cases have been recorded where from individual cells of, for instance, the leaf epidermis of *Begonia rex* or the hypocotyl of flax (*Linum usitatissimum*), a whole plant can grow.[1] Usually, however, this function is performed by sexual cells which divide in a special way after fertilization and give rise to the embryo in the seed.

The embryo is the primary source of the future plant, for, aside from one or more envelopes, it contains the rudiments of the root and stem. The directing forces that caused the arrangement of cells into these fundamental organs of the plant, the root and the stem, are not yet clear, if only because it is impossible to carry out experiments inside the seed as it develops on the mother plant without endangering its whole development. Studies of the development of an embryo are therefore more or less restricted to descriptions of the normal process of division of the fertilized egg cell, as shown in microscopic preparations. Such work has brought to light a great variety in the configuration (topography) of the partition process, but the deep causes of these variations remain concealed.

With regard to the agents that determine the differentiation of the embryo into the radicle and the plumule, we are also reduced to conjecture, based primarily on the nutritional requirements of these organs in the mature plant. Even experiments with young embryos extracted from developing seeds do not yield much information about the "correlations" [2] that determine the formation of the embryo. It is known only that the youngest embryos have very high requirements for organic substances, especially nitrogenous compounds, needed for the formation of protoplasm, while mineral nutrients and sugar often suffice for older embryos. An analogy with the apical meristem of the mature plant indicates that the totality of

[1] More recently this has been shown in those plant tissue cultures which break up into single cells, for on a number of occasions these cells have given rise to whole plants. (Ed.)

[2] The term "correlation" is used in plant development in a causal sense, i.e., to mean the influence of one part or organ in modifying the development or interrelation of others. (Ed.)

the plant is already determined in the earliest stages of its embryonic development, when all the cells are still capable of wide differentiation, and do not specialize till later, under special influences arising from other parts of the plant.

Much more amenable to these experiments are therefore apical meristems of stems, which can form a leaf typical of the species even if isolated from the plant and without direct connection through actual tissue with the older organs. This has been shown in a number of plants with sections only 0.1 mm thick (Wetmore and Morel, 1949). Probably correlations corresponding to the hereditary nature of a given plant are being observed here. An apical meristem that has been divided by three or four vertical incisions in order to break the contact with adjacent tissues and organs of the stem will form new leaves in normal arrangement (Wardlaw, 1952). Therefore it is in the apical meristem itself that the arrangement of leaves is determined; the same is basically true for their shape, although this can be altered under the influence of translocated material drawn from older parts of the plant. With a rich diet there comes an overcrowded arrangement in which one spot on the axis gives rise to more than two leaves.

At the tip are formed not only the leaves but also the buds, rudiments of future branches. With a well-placed cut, parallel to the surface, one can force the primordium of a lateral bud to change into a leaf, but the leaf primordium cannot be changed by such an operation into a bud. It follows that the leaf is the first structure to be irreversibly determined on the vegetative tip. This also holds for the tissues in a developing seed, where these first leaf formations gain priority over the other components of the embryo, that is, the radicle and the plumule.

Leaves are of the greatest significance to the morphogenesis of aerial parts of the plant. We learn this from experiments investigating the formation of buds, whose structure depends to a large extent on the influence of older parts. Woody plants are appropriate experimental objects, since each year they form winter buds. Buds were designated as long ago as 1790 by J. W. Goethe in his morphological studies as the primary sources of new individuals, and he emphasized the fact that in a bud one finds the rudiments of the future plant much better developed than in the seed. (Charles Darwin expressed himself in similar terms.) Experiments with buds may thus be especially important for the understanding of the future shoot as a whole and ultimately for understanding the whole tree. However, the experimental treatment must be applied to the buds very early, at the beginning of their embryonic development.

Primordia of new buds can be found in the lilac (*Syringa vulgaris*) even in very young rudiments of winter buds. They represent vegetative tips in the axils of small leaf primordia. These as yet extremely young bud primordia can be completely altered into leafy twigs, if we remove the small leaves from an unfolding bud. We can thus force young buds to grow into thin twigs; and if the leaves are removed, we find that only leaves develop at the lower end of these buds, where one would normally find scales as first indication of a new winter bud. Bud scales can appear only under the influence of green leaves, whose products confer on the base of the leaves the characteristic external shape and internal structure of scales. The scales are noticeable for their simple structure, the suberization of their surface cells, their brown color, and their greatly widened base. Following the first scales, new ones appear; in this way five pairs of scales generally precede the other components of the bud, that is, the young primordia of green leaves.

It has been possible to show experimentally that scales are not merely (as often thought) a protection for the delicate inner primordia of leaves and flowers against unfavorable mechanical agents and against dryness. Before they take up this function they act as important correlating organs, for they make possible the formation of real assimilating leaves. This has been proved by experiments with the large buds of the horse chestnut (*Aesculus hippocastanum*).

If we remove the first scales from a young bud just beginning to form, new scales have to develop before green leaves can be formed (Fig. 1b). If the operation is performed a little

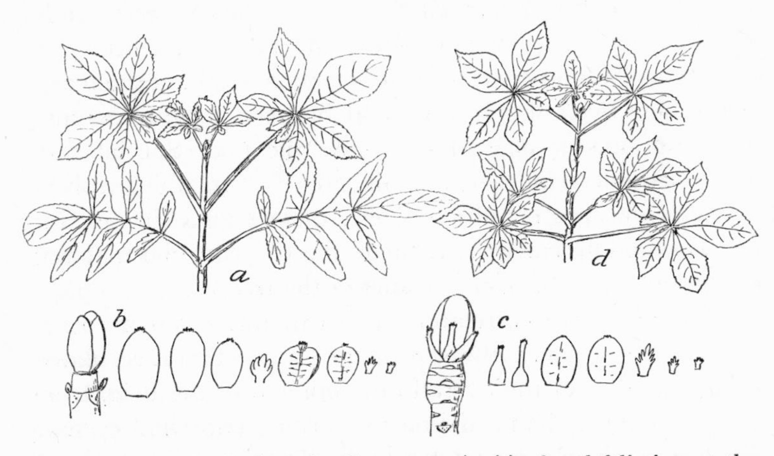

Fig. 1. Horse chestnut (*Aesculus hippocastanum*): (*a*) after defoliation, at the beginning of May, the new shoot which develops forms pinnate leaves, resembling those of the ancestors of the chestnut family; (*b*) following the amputation of two pairs of bud scales, two new pairs of scales are formed above the primordia of the first leaves; (*c*) when all the scales were removed, the first two pairs of leaf primordia were partially transformed into scales, then two new pairs of scales were formed, and only after these did leaf primordia appear; (*d*) the shoot which develops from the bud with amputated scales bears two pairs of scales between foliage leaves.

later, when the bud is slightly more developed and the first pair of green leaves has been irreversibly formed, then at first new scales will form in place of those that have been removed, and only thereafter will the rudiments of green leaves appear. We thus obtain a peculiarly arranged winter bud, in which scales are formed not only at the base but also in the middle, and distributed among the primordia of the leaves (Fig. 1c). The shoot that develops from such a bud in the following year carries scales situated between green leaves below and above, and separated by axial segments of the same length as those occurring between normal leaves (Fig. 1d).

The development of the future twig in the winter bud is therefore controlled by the leaves on the twig of the previous year, which determine the formation of scales in the rudiments of axillary and apical buds; these scales then make possible the appearance of green leaves inside the bud. There is a direct correlation between the leaves on a twig and the scales of the new buds of this twig, and between these scales and the primordia of new leaves inside the bud. Special substances produced in leaves or scales are very probably involved in these processes or, alternatively, these organs (in perhaps some quite different way) regulate the translocating material mobilized from the reserve tissues of the tree.

With these two correlations one can put together causal explanation of the rhythmic growth of a tree that is regulated by the rhythm of the overall development of the twig, from the beginning of its formation to its completion and even to the formation of new winter buds. The course of the embryonic differentiation of the winter bud is certainly also very important. This differentiation need not be the result of the influence of just one twig or even just one leaf of the twig on the primordia of the axillary buds. Even on big branches cut off from the rest of the tree, the formation of winter buds would be only partial if the branches were isolated early in the spring, before flowering. To complete the development of winter buds the action of substances coming from the roots is therefore needed. The structure of the bud depends, however, on its location, for it is influenced by many other organs that form the coherent whole of the plant. Experiments with fruit trees, like those performed by Tumanov and Garjejev (1951), are especially convincing in regard to the importance of flowers and fruits for the differentiation of future buds into vegetative or flowering branches. There are similar, but simpler, relationships in perennial herbaceous plants, and these lend themselves particularly well to experiments on the correlations between individual organs developing in the hibernating bud. Especially instructive are spring plants like hyacinth and tulip among the ornamentals, and *Ficaria verna* Huds. (of the Ranunculaceae) among the weeds.[3]

In Ficaria, which in its rigorous rhythmicity has some resemblance to trees, the buds for the next vegetative period are formed on tubers which develop in the spring of the year. Buds, however, form in these tubers only when the inhibiting action of the leaves decreases, that is, in the late summer and in the fall. In these buds one finds successively: — scales, primordia of foliage leaves, and (if the fresh weight of the tuber is at least 0.3 gm) a rudiment of a flower bud. All these develop at the expense of the food reserves of the tubers, and the number of tubers formed on the stock depends on the strength of the mother plant. The strength of the tuber stock itself will then determine whether the plant is going to have only a main stem, or also lateral branches, of which the apical ones bear flowers.

The number of leaves on the main stem, including flower primordia, is usually constant. Rigorous correlations are established between individual parts of buds from the very beginning of their embryonic development. For example, if we remove the scales from a bud before any leaf or flower primordia have become visible, then the primordia immediately following the scales develop directly into new scales, and if they have already been slightly differentiated into leaves, they are transformed into scales. The blade remains rudimentary, the base enlarges, and the petiole remains completely undeveloped. In *Aesculus hippocastanum*, as in *Ficaria verna*, scales have to be formed before any further development of the bud primordia is possible. If all the completely dif-

[3] Closely related to the American "Lesser celandine," or "Pilewort" *Ficaria ficaria* L. (Ed.)

ferentiated buds are removed (from robust groups of tubers), then new buds, identical to the first ones, develop from previously insignificant primordia on the stem or the tuber. In some cases this can be repeated several times.

Embryonic development of buds therefore uses up only a relatively small fraction of the reserve content of the tubers, so that much is left over for further elongation. Part of this is again used up in the second bud formation, if the first buds are removed. Up to a certain time, probably, the composition of the reserve content does not substantially change, so that new buds can still be formed. This behavior shows that the differentiation of the bud into scale, leaf, and flower primordia cannot be due to progressive changes in the composition of the reserve substances of the tubers, but must be a product of the correlations established between the various parts of the bud as a whole.

It is only when the bud is allowed to develop past the time when the first foliage leaves grow out that the composition of reserve substances changes so profoundly that the new buds, which develop upon the removal of the original ones, bear a great many more scales; sometimes they even comprise nothing but scales; as a consequence the initiation of leaves is inhibited. Such buds, along with many other bud rudiments, remain closed in the following spring.

Under the influence of the growing leaves, inhibitions develop in the reserve materials to such an extent that only scales can be formed, the further production of leaves being limited or totally inhibited. Tuberous branches in which this stage has been reached have to go through a rest period in order to be able to form leaves in their buds; as a result they live one year longer. The regulatory influence of scales appears very clearly from this experiment. The plant is rejuvenated, since between the two phases of bud formation there has been a rest period.

It is impossible to perform the same experiments with

seeds, even those which contain amounts of reserve substances comparable to those in tubers. For example, in the seeds of the pea, *Pisum sativum*, the cotyledons are followed by two primary scales, marked by a wide base with three dents, then by two folded leaves each with a stipule, a petiole, a pair of leaflets, and a tendril. This differentiation into various leaf components is governed by correlations established in the seed. In this case the cotyledons can be compared to the foliage leaves of woody plants. Both primary scales arise under the inhibitory influence of the cotyledons and are necessary for the formation of typical foliage leaves, which cannot follow directly after the cotyledons with their rich reserve content of proteins and starch.

Goebel (1880) refers to scales as inhibitory structures. However, it would not be right to consider them as leaves inhibited in their growth, for, as has been recently pointed out (Foster, 1929), the scales represent formations of a special kind. Foster expressed his surprise at the fact that scales are found in seeds, since the seeds represent the foundation of the future plant, which rejuvenates through this mechanism. Krenke (1950), in his theory of cyclic aging and rejuvenation, stresses mainly multiplication through seeds. Any other explanation of these scales on the pea embryo would undoubtedly encounter difficulties.

It should be noticed that the internodes elongate between the primary scales of the young seedling of the garden pea, as well as between the scales situated among foliage leaves on the branches of the horse chestnut. As a result, in both cases the scales are not crowded upon one another, as they are normally at the base of winter buds. At the same time we know that the two lowest internodes, separating the cotyledons from the scales, differ from the other axial segments (which only separate leaves from one another), by their reaction to the unilateral influence of light or gravity. Their positive phototropic response is stronger than their negative geotropic one, so that they can bend only toward the light. However, in the third internode, below the first green leaf, positive phototropism is suppressed by negative geotropism, so that the stem straightens up, and even turns away from light. If we illuminate embryonic pea stems from underneath, they bend downwards when they have only the first two axial segments, but as soon as their third internode reaches a certain size they straighten up again.

In the seed, when the rudiments of the main stem (the plumule) have been formed, buds develop in the axils of the cotyledons. Each axil bears a whole row of these buds, so that they could replace the main stem several times if it is damaged. It is clear that these secondary primordia have been inhibited in their development by the cotyledons and plumule. In the seed, correlations already exist between the apical and the axillary buds — the so-called Apical Dominance — which is the most important principle regulating the arrangement of aerial organs on the plant. The apical meristem of the embryonic shoot, probably under the influence of the cotyledons, gains a superiority over the lateral buds, and first of all over the cotyledonary buds. This correlative superiority can be traced all the way to the seed. It is for this reason that lateral buds remain quite small; only when the inhibition is lowered, following the formation of the two primary scales on the plumule, do leaf rudiments start to differentiate in the lateral buds. These differentiate not as scales but as foliage leaves, in phase with the leaves which are being formed at the same time on the main axis above the two primary scales.

The same course of development can occur in cotyledonary buds, if during embryonic growth their previously undifferentiated leaf rudiments finally differentiate. Such action, which begins with compound foliage leaves, can best be demonstrated by removing the main stem and any subsequent shoot tending to replace it. However, if this is done only after the shoot has reached a certain size, typical scales develop, before the leaves on the last lateral shoots. They are very similar to the primary scales which are already formed on the plumule in the seed. During development, as the embryonic pea plants seem to show, there is an increase of the inhibitory influences in the basal part of the plant, reaching a level comparable to that present when the embryo begins to form in the seed. It must be emphasized that such inhibitions act specifically on those organs which start to differentiate during the time of their greatest activity. Because the scales on the last aerial shoots are followed again by typical foliage leaves, scales can be assigned an important regulatory role, namely that of reducing the inhibitions in the plant.

The development of leaves is a visible sign of the condition of the plant at the time; experimental morphology provides many examples of this. The shape of the foliage leaves of the whole plant is largely determined in the bud and so also is its anatomical configuration; this was first demonstrated on light and shade leaves formed on branches of the same tree, some of which were exposed to direct light and others shaded. Small rudiments of leaves, along the whole length of the future stem, are already irreversibly determined in the bud. At the base of stems and at the initiation of new growth, inhibitions obviously decrease and ultimately increase again. These differences are clearly shown in the extensive studies of Krenke, which supplied the basis for his theory of aging and rejuvenation. From the rapidity of these changes one can judge the economic properties of both herbaceous and woody cultured plants. They have also practical importance for the prediction of the flowering season, the size of the crop, and the resistance of cultured plants.

Another interesting example is provided by the experiments of Ashby (1950) with *Ipomaea coerulea*. This plant, if grown under short daylight conditions, forms simple leaves along almost all the stem; in long days it forms simple leaves only at the base and at the top of the stem, and compound leaves in the middle. Nevertheless, this shape is irreversibly determined in the bud, when the initials are still very small. Here again the integration of the plant is clearly shown, since a certain shape of leaves is determined on the whole stem at the same time and affects buds at various levels and various stages of development.

Only leaf primordia of a certain age can react to the metabolic change which brings about a definite leaf shape. It is

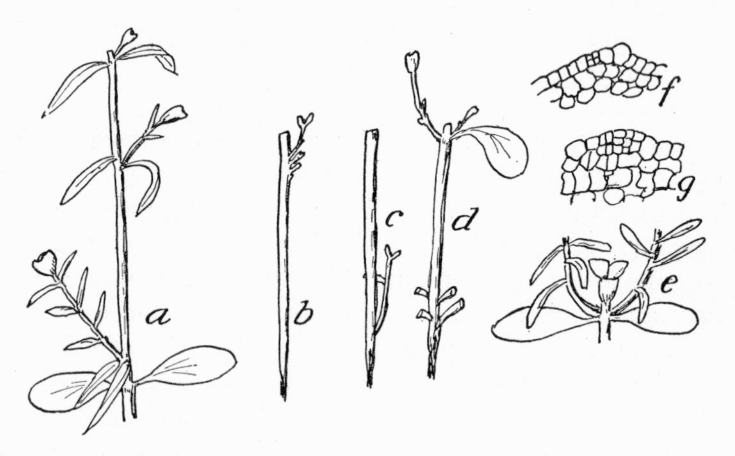

Fig. 2. Flax (*Linum usitatissimum*): (*a*) when triiodobenzoic acid paste is applied on one of the cotyledons and the cotyledonary buds amputated, together with the tip of the stem, the shoots which develop from ring-shaped leaves at their tips — this occurs at the bottom of the stem, after six normal leaves; in the middle of the stem, after two normal leaves; and at the top of the stem, directly from the first leaf primordia which fuse together; (*b*) a young hypocotyl forms adventitious buds at the top; (*c*) an older hypocotyl forms adventitious buds at the base; (*d*) after application of triiodobenzoic acid paste on the cotyledon, the bud in the axil of the amputated cotyledon grows more than the one opposite. At the same time adventitious buds form at the base of the hypocotyl; all of these buds form ring-shaped leaves; (*e*) when seeds are soaked in a solution of triiodobenzoic acid, an ascidium forms instead of the first pair of leaves. Later, however, normal cotyledonary shoots develop; (*f*) section through the external layers of the hypocotyl, showing an epidermal cell which begins dividing and gives rise to adventitious buds; (*g*) the same, later.

for this reason that in the buds situated lower down on the stem the leaf shape appears closer to that of the upper part of the bud meristem than it does to the buds situated higher on the stem. This change can be compared to the changes in shape caused by certain growth substances, such as 2,3,-5-triiodobenzoic acid, which spreads from the point of application to the rest of the plant. This substance, foreign to the plant, in some cases brings about an increase of flower production, in others a fusion of leaves, by inhibiting the elongation of axial segments between them. The latter reaction occurs in flax, where in the embryonic stage the leaf rudiments of a certain age on treated plants undergo abnormal fusions, so that three or four leaves are fused together to form a funnel-shaped ascidium (Fig. 2a). This can be demonstrated by removing the apical bud of the young embryo above the sixth or eighth leaf, and at the same time by dissecting out the buds in the axils of both cotyledons, for they would be more likely to grow than the lateral buds on the decapitated plant. If now one of the cotyledons is treated with lanolin paste containing 0.5 percent triiodobenzoic acid, all the remaining primordia start to grow, because this substance induces branch formation. At the time of application the various primordia along the stem are at different stages of development; the lower ones are much more differentiated than the higher ones. The funnel-shaped fusions appear therefore only on the branchlets which grow from the axils of the first leaves, right above the cotyledons after six normal leaves; on the branchlets growing out from the middle nodes ascidia follow two or three simple leaves, while on the top shoots there are no free leaves at all and the ascidium grows straight from the axil. Hence only those few leaf primordia which happen to be at a particular stage of development can respond to this stimulus and fuse. Those which are either younger or older develop into free alternate leaves, separated from one another by normal internodes. In other words, though the substance has penetrated the whole plant, only certain parts react to it and others develop normally.

The influence of triiodobenzoic acid is similar to that of the so-called flowering stimulus, which is just a way of referring to the unknown chemical mechanism responsible for the development of flowers. Probably this stimulus influences the whole plant, although only specific elements react. Thus in the lilac, flower buds are formed only on the topmost nodes of the plant; on the lower nodes only leaves appear, because the formation of flowers there is prevented by strong inhibitions.

The nature of the flower bud of the lilac is determined very early, at the end of April or the beginning of May, when the upper part of the plant is not yet completely formed. The buds are then quite small, bearing only the outer scales. The inner scales are tiny and surround a still flat, central part, with only a few indications of leaves. Careful observation of the lowermost of these buds in the fall shows four or five pairs of scales. In the buds situated immediately above these, the number of sterile scales decreases to two pairs or even one pair, after which follows a pair of bracts with visible primordia of flower buds.

These sterile scales situated below the floral parts of the lilac have a marked resemblance to the lower unfused leaves of flax seedlings treated with triiodobenzoic acid. When bud primordia on the top of the plant are inhibited in their growth to such an extent that even scales cannot be formed in sufficient numbers, they remain forever vegetative even though they may be situated very close to flower buds or even above them. When later in the season the inhibitions decrease, leaves rather than flowers form in these buds because by then the flowering stimulus is inactive.

Similarly in fruit trees, an overproduction of flowers and fruits inhibits the differentiation of new buds; but after the fruits have ripened and these inhibitions have decreased, the formation of flower buds is no longer possible and only foliage leaves develop. The main cause of the periodicity of fruit production, however, is the strong inhibition originating in the leaves, as in the case of the lilac.

Bud differentiation thus depends on many external and internal factors which determine the future development of the shoot and of the whole plant. Inside the bud there are correlations similar to those occurring later between individual growing parts of the plant. It is amazing that in the bud, within such a tiny space, such marked differences between individual primordia can exist. The interactions between them are independent of the functions that the organs developing from these primordia will possess. In this respect plants, especially the most highly organized ones, are probably very similar to animals. The organic unity of the plant as a whole, which is prepared so far ahead, is obviously a result of phylogenetic development. In every stage of individual development the plant presents an integrated whole governed by reciprocal interactions between its various organs.

There are many technical obstacles to the study of the correlations governing the embryonic formation of buds. But of the many examples, the importance of scales stands out clearly. The scales act as mediators: on the one hand between the inhibitory influences of the green leaves and the cotyledonary reserves, and on the other hand between the new formation of leaves and that of other organs. In plants which do not form scales, at first we find leaves with very simple shapes, which can be considered as atavistic forms. It seems that scales enable plants to drop these ancient shapes and to come directly to the present form of their morphogenesis. This particular role of the scales is demonstrated when phylogenetic recapitulation is induced experimentally.

CHAPTER II

Experiments on the Recapitulation of Phylogeny

A valuable guide to the study of the causes of individual development (ontogeny) is Michurin's basic law of the unity of ontogeny and phylogeny.[1] Phylogeny represents the sum of all the ontogenies that have gone before and have been modified by external influences of the environment. Through continuous repetition they have gradually become concealed in contemporary organisms, but traces of them often appear at the beginning of development as ancestral forms (atavisms or retentions). It has been proved, especially in animals, by innumerable anatomical and embryological studies (among which should be mentioned the extensive work of the Russian anatomist and embryologist Severcov, 1911) that during early ontogeny, phylogeny is briefly repeated.

Naturally it is impossible to experiment with the past development of species, that is, with phylogeny, because one would have to work mainly with forms that have died out and of which only the remains are preserved in the various layers of the crust of the earth. It is nevertheless possible to follow

[1] Or, as Haeckel (1860) put it long before Michurin, "the ontogeny repeats the phylogeny," i.e., the individual in its development follows the changes which the species underwent in the course of its evolution. (Ed.)

phylogeny experimentally in the first embryonic steps in which it is partially repeated, if suitable experimental objects are available. For such experiments plants are certainly more suitable than animals, since the enclosed development of animals and their limited power of regeneration, especially at the higher levels of organization, do not permit much experimentation. Even so, the more primitive animals with better regenerative power, such as the Molluscs and Echinoderms, show, after the loss of certain parts of their body, new regenerated parts reminiscent of their ancestors. If a given form is the necessary result of a particular metabolism, then the regenerations must have been carried out at the expense of products which entered into the composition of the ancestors of the regenerating animals; thus the ancestral form was able to combine with the modern recent one to form an integrated whole.

Occasionally such cases are also found in plants, where systematists interpret them as "retentions." R. von Wettstein (1933), the Viennese botanist, cites F. Müller (1864) for whom ontogeny is a short and fast recapitulation of phylogeny. Haeckel (1860) expressed in similar terms his so-called basic biogenetic law. According to von Wettstein, however, one cannot speak of a simple recapitulation of phylogeny through ontogeny, for he accounts for changes by mutations, as a result of which there arises something new that could not have been inherited. He claims also that adaptation plays a role in evolution, so that only the forms that do not undergo the pressure of external conditions could have persisted. On the other hand, he admits that adaptation may be atavistically repeated in an early developmental stage; in this way he also explains the juvenile shapes of leaves in embryonic plants, if they differ from the shape of leaves on the adult plants. The pinnate leaves that arise on seedling vetch plants (*Lathyrus aphaca*), following the two primary scales, resemble original forms of vetch plants which are found in more humid locations and which still have more or less pinnate leaves. They thus represent atavistic remains from the times when even this species inhabited more humid places before transferring to dry land and forming enlarged phyllodes without blade-bearing leaflets, instead of true leaves.

According to von Wettstein and many other systematists and phylogeneticists, it is impossible to obtain information about the phylogeny of plants by experimenting on them. On the other hand, the greatest Russian Darwinian, Timiriazev (1922), in his "Historical Method in Biology," designated just this topic, experimental morphology, as the basis for the study of evolution. In reality the more we discover about the causes of development in today's plants, using experiments at all stages, the better we shall understand their phylogeny.

It must be realized, however, that the evolutionary factor, or a purely historical approach to the development of organisms, represents only one point of view from which to approach evolution, and our conclusions from this alone would be oversimplified if they were not complemented by analysis of the causes. The same thing can be said about the gene theory, which does not clarify the development of form, neither of itself nor from the point of view of ontogeny or, least of all, from that of phylogeny. For between the hypothesized gene and the actual realization of forms under the influence of external environment lies an as yet unbridged abyss. One cannot satisfactorily explain, from the point of view of the gene theory, the phylogenetic recapitulations obtained experimentally by removing leaves or by such other techniques as treatment with specific photoperiods or with artificial inhibitors.

If, at the beginning of May, we remove all the leaves from one branch of a large horse chestnut tree when the first leaves are starting to mature and the new buds are still small, these buds, which are still at the beginning of their differentiation as winter buds, develop into branches bearing pinnate leaves instead of their normal palmate ones. These will then be followed in the upper parts by the regular palmately compound leaves (Fig. 1a). The same thing can be seen in a smaller tree if we defoliate it completely. These changes cannot be explained by mutation, since no horse chestnut with such pinnate leaves is known. Also, the pinnate shape is not retained after defoliation. The experiment is successful only if performed in the first half of May. If branches are defoliated later, when the development of winter buds has progressed further, the leaves that develop on the new branches are only palmate.

If we defoliate the branches bearing pinnate leaves, brought about by defoliation in May, then from the yet unformed buds there develop new branches with only typical palmate leaves. The pinnate leaves are therefore related to a particular state of the plant, marked by an altered metabolism that in the first half of May specifically corresponds to the metabolism which existed in this species of plants at the time when it formed only pinnate leaves. (The same type of pinnate leaves is found occasionally in nature. They result probably from the metabolic conditions caused by exceptional mechanical influences: entangling of branches, loss of leaves through frost, and so on.) This has indeed not escaped the attention of teratologists. Penzig (1921) compares them with leaves of a similar shape found in the Sapindaceae, a related, mostly tropical, family from which botanists derive the horse chestnut family Hippocastanaceae. One can hardly be dealing here with adaptation in von Wettstein's sense, for both shapes of leaves — pinnate or palmate — fulfill the same functions, that is photosynthesis, transpiration, et cetera.

These morphological differences are much more clearly understood in terms of the biogenetic law. For we can explain pinnate leaves as ancestral forms carried over from common predecessors of both of these families. It is worth while to stress that the formation of these ancestral leaves can be brought about in the horse chestnut experimentally with absolute regularity. This suggests the possibility of the *experimental discovery of ancestors* of today's plants, through interference with embryonic development before the final forms of the modern shapes have become established. At the same time, the metabolic requirements for intensive growth by elongation must be satisfied, so that this ancestral form can be not only induced but also fully developed. If this basic requirement is missing, only normal scale leaves will be formed and following these (with no transition) modern-shaped leaves will appear.

There are comparable responses in other cases where the first stages of embryonic development have been interfered with by defoliation. For example, the European ash tree, *Fraxinus excelsior*, if defoliated early in the season, forms new branches beginning with *simple* leaves, followed later by the normal pseudopinnate leaves. In this species, one can obtain simple leaves only on the top branches; when the lower ones are defoliated, the new leaves are only pinnate, probably because of greater inhibitory influences coming from the crown.

The occurrence of simple leaves in the ash has been given various interpretations. The one most commonly accepted is that simple leaves represent the original form. This form has survived in certain species such as *Fraxinus monophylla* or *F. diversifolia*, although H. de Vries (1906) considers these to represent mutations.

Neither these simple leaves of the ash nor the feathery leaves of the horse chestnut can be considered as modified scales, for scales with their broad winglike bases and with no trace of a petiole are formed differently from the outset. Those ancestral simple leaves have, however, in common with bud scales, a regulatory function, for in the same way as do scales they allow the formation of primordia of typical modern leaves, even though conditions remain favorable for the continued formation of the simple ancestral type. Sladky's (1957) experiments show this very clearly. At the beginning of spring he removed the leaves of the top branches of some large ash trees and found that two pairs of ancestral leaves appeared on the new branches, which bore a small number of scales; thus it would appear that each pair of ancestral leaves must correspond to the formation of a great many pairs of bud scales.

Ancestral leaves therefore replace scales, and this phenomenon may offer an alternative phylogenetic explanation for the nature of scales. Using the explanation of embryologists, who have concluded that from among the oldest phylogenetic characters those that are absolutely indispensable for life have been preserved, we can say that here the scales were preserved because of their great importance in protecting the buds from unfavorable conditions. They are therefore especially common on woody plants of cold and temperate latitudes. Their appearance on some trees of warm and humid tropical climates, however, indicates their primary regulatory significance, which functions even in constant climates on periodically growing plants.

The fact that *Fraxinus excelsior* forms the ancestral simple leaves more easily after defoliation of the top of the tree than of its lower branches (which are more under the influence of inhibitions coming from the crown) demonstrates the importance of integration in plants. For even these inhibitory recapitulations which occur at the apex are directly helped by the root system. The defoliation of bottom branches, which does not lead to ancestral leaves, does permit the formation of a greater number of scales and later recent pseudopinnate leaves can be directly formed.

Some of these ancestral forms might be considered, however, as modified scales, or at least as transition forms between scales and typical foliage leaves. An example of this is found in the lilac, *Syringa vulgaris*, a plant very convenient for experimentation because of its opposite leaves. If we defoliate lilac branches in the spring, or if we cut off the tip — a common practice for the shaping of summer lilac hedges — then we observe new shoots, arising from the primordia of winter buds, and at first bearing leaves of strange shape that resemble those of *Ginkgo biloba* (Fig. 3a). This Gymnosperm is, how-

Fig. 3. Lilac (*Syringa vulgaris*): (*a*) the first leaves formed on the shoot which grows after defoliation have a highly branched venation, somewhat resembling that of *Ginkgo biloba*. The shoot is terminated by a single bud (monopodium); (*b*) two middle pairs of leaves on a strong shoot: the lower one has branches terminated by a pair of buds (sympodium), the upper one by a single bud (monopodium).

ever, very far from the lilac and the similarity between leaves is only superficial; one cannot draw any conclusion as to a real phylogenetic relationship. In any case *Ginkgo biloba*, the leaves of which have undergone significant changes on the extinct forms, regularly bears leaves of different shapes on long and short shoots, and the above-mentioned ancestral forms of the lilac are closest to the leaves on the brachyblasts (short shoots), which also represent delayed forms. These leaves are rather broadly wedge-shaped with a typical palmate venation, with veins branching out directly from the base. In contrast, the venation of normal lilac leaves is pinnate. To this extent these ancestral leaves do resemble scales, for in scales also a great number of veins originate directly from the base. We might conclude, therefore, that the original scale primordia, after the loss of leaves, have been exposed to metabolic forces which cause the formation of normal leaves. This interpretation is supported by Goebel's observation (1880) that similar shapes arise when scales become modified; as already mentioned, he regarded scales as only inhibited forms of leaves. However, the pinnate leaves produced on the horse chestnut, so radically different from the typical palmately compound ones, show that the effect of defoliation is to bring about the formation of more primitive ancestral shapes of leaves, and thus make it probable that the abnormal leaves produced on lilac are also ancestral.

Equally informative is the arrangement of branches in the lilac, which represents a sort of transition between monopodial and sympodial branching. Such an arrangement can be seen in many other plants. In the monopodial type, the main axis grows continuously and forms weaker lateral branches, while in the sympodial the apex of the main stem degenerates and is replaced by a lateral branch. It is possible to prove that in the lilac monopodial branching is more primitive and linked to certain inhibitions, while sympodial branching is more advanced and caused by specific stimulations. It is generally recognized that sympodial branching is the more evolved type. It leads to the formation of a fuller crown of the tree, to a greater development of the photosynthetic area, and therefore to a greater fruitfulness. It is present in today's flora in something like 90 percent of our woody plants.

If we defoliate or prune spring branches of lilac at a time when the winter buds can still grow into branches, the new "proleptic" shoots, which are always much shorter than the first spring branches, show a regular monopodial ending. In other sympodially branching bushes (that is, where the tip of the branch regularly stops growing, dies, and is replaced by one of the lateral buds), the most apical leaves in the bud are preserved but are inhibited in their growth after defoliation. The basal parts are transformed into scales and their blades wither away. Only then do true scales appear. Sometimes this happens in the lilac spontaneously, when the metabolic conditions are such that the latest primordia of green leaves change at their bases into scales, while the true scales with no blades are developed afterwards.

These changes can also occur in axillary shoots which appear about halfway up the stem of very strong basal shoots; both in the lower and in the upper parts of the stem the axillary buds remain completely inhibited. These so-called "sylleptic sprouts" mark the position of most active growth as being in the middle of such vigorous plants. If two pairs of such axillary shoots develop, then each of the lower pair ends in two lateral buds (which means that the apical bud has died) while each of the upper pair is terminated monopodially by only a single bud (Fig. 3b). The upper part of the stem thus shows a different phylogenetic nature from the lower part. This agrees with Lysenko's experiments (1952, 1954) on the soybean (*Glycine Max*), which indicate that the two parts reach maturity at different stages: the lower part of the stem is stagewise younger than the upper part. We can correspondingly consider the lower part of our new lilac shoot as phylogenetically younger and the top part as phylogenetically older.

There is a gradient of inhibition from the roots to the top of the tree. The above-mentioned phylogenetic ages or stages of maturity, expressed in the production of flowers, correspond to this gradient. All this undoubtedly depends upon metabolic differences existing between various parts of the stem. Similarly in the case of galls, caused often in the lilac by the mite *Eriophyes loewi*, the apical buds are preserved and the growth becomes monopodial, so that the parasite appears to produce the same metabolism as the defoliation experiment does on normal, healthy bushes.

Similar differences in branching are also found in other woody plants of sympodial habit. If the experiment is carried out at the right time of the year, the terminal buds will always be preserved and will develop into leaves which, according to all results, represent primitive shapes. A suitable experimental object for this is the linden tree (*Tilia*), interesting for its striking change in leaf shape (Fig. 4). For

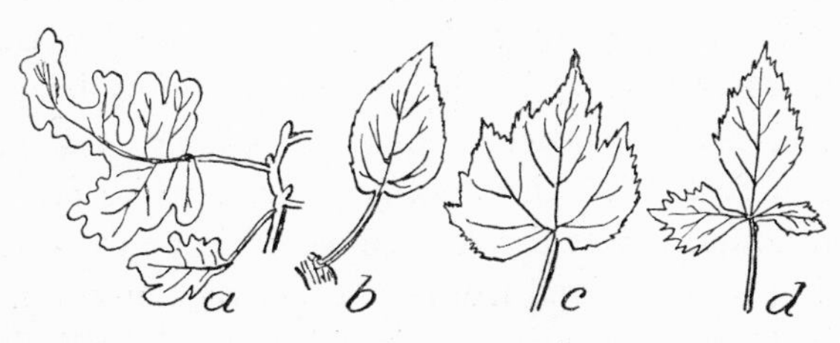

Fig. 4. Linden (*Tilia platyphyllos*): (*a*) lobed leaves ("forma *laciniata*"); (*b*) normal leaves from the trunk; (*c* and *d*) new leaves, following defoliation of the normal linden (after J. Vaněk).

example, *Tilia platyphyllos f. laciniata* normally has laciniated, irregularly lobed leaves, but if a shorter shoot grows on a thick branch or directly on the trunk, it forms normal, heart-shaped leaves, with only a toothed margin. In the same way, very short shoots, growing spontaneously on strong branches of lilac, remain monopodial. The remarkable thing, however, is that after the removal of leaves on normal (non-laciniated) linden trees, there arise, on both sides of the tip, leaves which are more or less lobed or even palmately compound; this happens only if the cut reaches the insertion of the blade on the petiole. Again the phenomenon can be ascribed to an inhibition acting on the very young bud primordia, which are not protected by a sufficient number of scales, for these have been precociously forced to elongate into the ancestral form. This form appears even in the normal cleft cotyledons of several varieties of the linden tree. Usually the form of cotyledons on embryonic plants is simpler than that of foliage leaves, but in the linden tree it is the other way around.

We cannot conclude, therefore, that the simple blade is more primitive than the compound one. Indeed, the presence of very complicated compound blades on ancient types of vascular plants like the ferns is sufficient indication of that fact. If we defoliate the typical linden tree, there are formed, instead of simple toothed leaves, shoots carrying more or less lobed leaves; on the other hand, when a linden tree with compound leaves is defoliated the new shoots bear simple leaves. This indicates that in the normal linden tree the compound leaf represents the ancestral form, while for the tree with compound leaves the reverse would have to be the case, one having been derived from the other by mutation.

In other laciniated forms, as in this case of the linden tree, we can also observe reversion to the more primitive simple leaves. Such reversion corresponds to an increase in inhibition. A good example is the laciniated variety of the winter oak (*Quercus petraea*). This tree forms laciniated leaves on the first spring shoots, while the shoots which come later, growing out from the tip of the spring shoots at the end of June around the time of the feast of St. John, show simpler leaves with only slight lobes, corresponding to the typical original form of this oak (Fig. 5). Even here we can detect inhibition gradually increasing from the roots to the tip. Here also the April shoots are, both phylogenetically and stagewise, younger than the June ones.

The existence of these inhibitions is shown also by the fact that a great many branches of big trees do not produce such summer twigs, so that they can often be found only on the thickest branches. Small trees, especially if they are limited in water supply, produce summer shoots only rarely, while heavily watered trees, with a highly developed root system, may burgeon twice or even three times a year. In this case the integrative forces in the plant are very evident, because

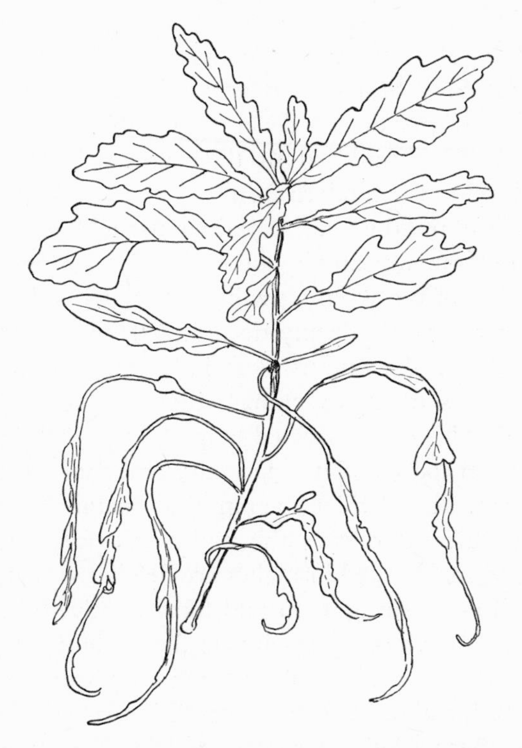

Fig. 5. Oak (*Quercus petraea*, forma *laciniata*): summer shoot, developing above the spring shoot, reveals the original form by its normal leaves (after Späth).

the rhythmical growth of shoots is helped by the development of the root system. For this reason, on branches growing from stumps one finds several series of summer branches — a result of the dominance of the original, large root system, which remains active although the tree has been cut down.

Defoliation or deep decapitation brings new growth to an area in which growth had previously stopped at a point where the future buds were not completely formed. This operation can therefore be considered a reliable method for obtaining the growth of more primitive, or of ancestral, forms. Normal embryonic development of modern-shaped organs can be inhibited by the fact that the operation brings about the right conditions for intense elongation. If this does not happen the ancestral forms do not develop, because the primordia of the new organs, inhibited in their growth, will thus gain time to develop the modern form.

With this in mind we can explain a case which at first seems to contradict our working hypothesis. This concerns the hanging branches of the weeping willow, *Salix viminalis f. pendula*. If these branches are deeply decapitated, some of the shoots developing from the stumps (especially of big branches) grow straight up, as on nonweeping willows, instead of arching downward. This was described long ago by Vöchting (1884). The first internodes on later, arching, branches are sometimes still vertical. It is evident again that metabolic conditions have been altered during the embryonic formation of the shoot. If normal arched branches are defoliated, then the new shoots will grow straight up and form relatively short internodes; this behavior again points to the straight growth as the primary form from which the arched type arose secondarily. We can bring about the same conditions experimentally in other weeping varieties, for example in the ash tree (*Fraxinus excelsior*, *f. pendula*). The prerequisite for the formation of normal straight branches is here an incomplete embryonic development of buds, together with a vigorous root system which favors precocious elongation.

Within the living matter of today's plants must be concealed the power to reveal their ancestors, under the proper conditions. Ancestral forms can appear, at least transitorily, before conditions revert to normal and the modern forms develop again. As an example, the noncellular alga, *Caulerpa prolifera*, usually produces green "leaves" or assimilators, elongated, egg-shaped or even heart-shaped, similar to those of higher plants. But if this plant is forced to grow during its usual rest period, as for instance when it is transplanted from the sea to the laboratory between January and March, it produces from the remaining rootstocks and leaf bases thin, round, delicately branched structures resembling the filamentous branched assimilators of the simplest species of the family *Caulerpa fastigiata* found in warmer seas (Figs. 6 and 7). The

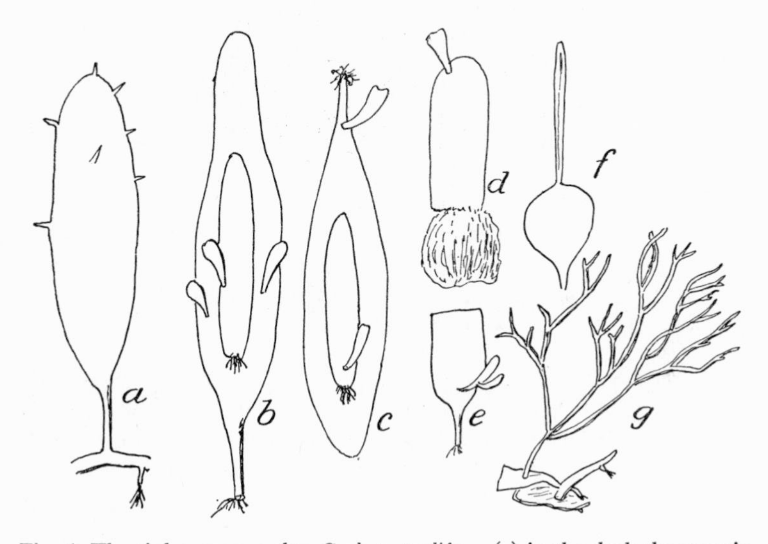

Fig. 6. The siphonaceous alga *Caulerpa prolifera:* (*a*) in the dark the margin of the assimilator loses its inhibitory effect on the formation of new assimilators, which then grow out from the edges; (*b*) an elliptical section separates the outer and inner parts of the assimilator; both parts, when normally oriented, develop rhizoids at their base; (*c*) in the inverse position, the outer part keeps its original polarity, while the inner part reverses it; (*d*) the upper part of the assimilator, when grown in mud, forms many rhizoids and branches immediately below the tip; (*e*) the basal half, floating freely in water, forms few rhizoids and shows basal polarity; (*f*) the new blade developing from the previously inhibited apical meristem is perpendicular to the original blade; (*g*) in January, during the rest period, primitive, filamentous forms of assimilators appear.

same shape appears under normal culture conditions in non-resting plants, if the temperature of the water is allowed to rise; presumably this brings about an increase in inhibitions. Similarly, only threadlike "leaves" (assimilators) are formed even on normal plants during the hot months, especially in a culture without mud, obviously because of very high inhibi-

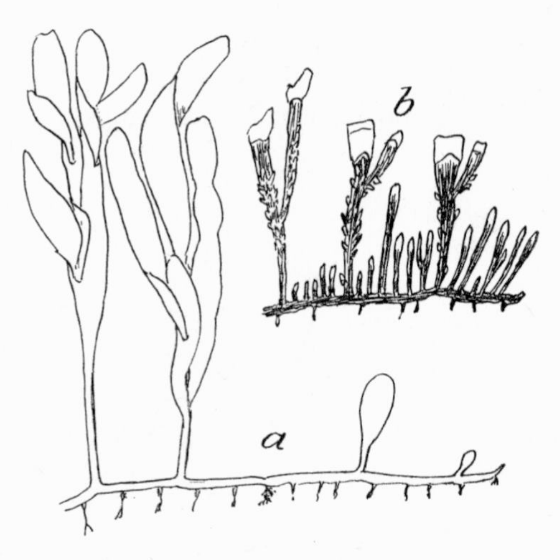

Fig. 7. The alga *Caulerpa prolifera:* (*a*) normal plant, showing regular development of assimilators and rhizoids at the apical end of the rootstock; (*b*) under the influence of running water, the green content of the assimilators is transported into the lower parts, especially into the rootstock, which, if restored to calm waters, shows a chaotic development of narrow shoots.

tions. Under such conditions this noncellular plant reverts to primitive forms like those of the large group of noncellular algae in the family Siphoneae, which are quite close to the feathery algae of the genus Bryopsis or to such simple filamentous species of the genus Codium as *Codium dichotomum*. If we improve the culture conditions, for example, if the rest period is over, or the temperature of the water is lowered, or it is enriched with the organic residues contained in mud, then

Caulerpa prolifera develops its normal leaflike assimilators. However, in running water cultures it remains in the primitive stage, with all the living substance transferred to the rootstock and to the basal part of the assimilators, which for the most part are emptied and die. Here, therefore, all the living matter migrates out of the main regulatory organs, or we could say that the "leaves," comparable to leaves or cotyledons of higher plants, take refuge in the rootstock, and this causes the plant to revert to its oldest forms (Fig. 7b).

These results suggest that the inhibitions in a plant can be increased in various ways and then can be used for phylogenetic recapitulation. In long-day plants the inhibitions are increased by short-day photoperiods. In short-day plants increased inhibitions favor flowering. Even in long-day plants flowering requires a definite, although relatively low, level of inhibition of vegetative growth. However, (some) long-day plants grown in short-day conditions do not lengthen their axial internodes, but form rosettelike structures. For this reason they do not flower, but this can be corrected with gibberellic acid.

Among the plants suitable for such experiments is the spring crowfoot, Ficaria, a member of the Ranunculaceae, belonging to the oldest group of dicotyledons, classified together under the name Ranales. If this plant is grown during short winter days at room temperature, so that the bud primordia formed the previous autumn can start to elongate, certain anomalies appear in their growth, somewhat like those already mentioned for the marine alga *Caulerpa prolifera.* Since all the organs of the future plant are already preformed in the bud, we cannot expect any drastic change in this respect, even under winter conditions. Only the roots, which are usually the most plastic organs of a plant, will undergo noticeable changes under the influence of short-day photoperiods and an altered rhythmicity of the plant.

Ficaria possesses on the one hand typically threadlike and

highly branched absorption roots and, on the other, egg-shaped, club-shaped, or even spherical reserve roots; these latter are rounded off at the tips and full of food materials, especially starch, accumulated in the primary cortex. However, during winter growth, at least the oldest of these tuber-like structures start to elongate, and from the cylindrical basal part grows a long-branched absorption root (Fig. 8d). This

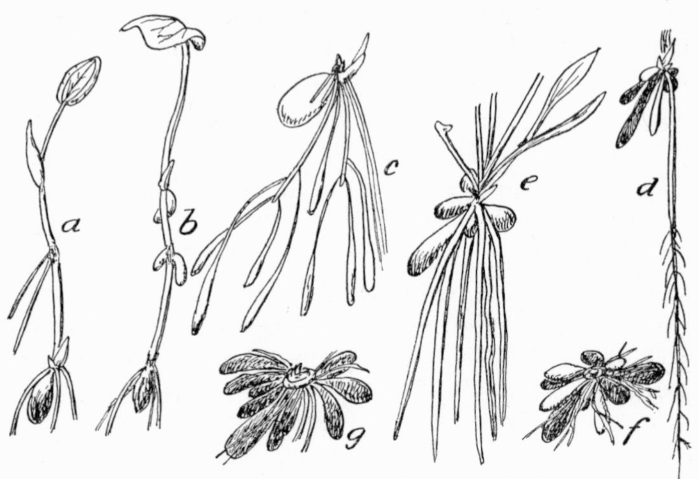

Fig. 8. Lesser celandine (*Ficaria verna*): (*a*) when plain lanolin paste is applied to the middle of an elongated rhizome of a weak individual, roots are formed; (*b*) when auxin paste is applied on the same site, tubers arise; (*c*) under continuous electric light, root tips become transformed into tubers; (*d*) during short winter days the first tubers develop into roots; (*e*) following the amputation of the stem, the formation of roots is replaced by that of elongated tubers and the formation of kidney-shaped leaves by that of comma-shaped ones; (*f*) after removal of the entire stem and application of auxin on the rhizome, new tubers are formed; (*g*) on control plants, treated with plain lanolin, only weak roots are formed.

is similar to the spherical root of the sugar beet with its broad basal reserve portion and a thin region active in absorption.

If we amputate the normal absorption roots at an early stage the elongation of the tuber roots above them becomes accelerated and they develop a more rootlike form. Inhibitions are increased in the plant, a phenomenon which commonly follows the loss of roots. Among other things, the movement of leaves serves as a sign of these inhibitions. In weak light when the inhibitions are relatively weak, the leaves point upward, while in strong light, where inhibitions increase, they expand horizontally or bend down. This epinasty is here, as in other plants, promoted by the removal of roots, provided that the plant still contains plenty of reserve materials. A great number of different shapes of tubers can be obtained in this plant by appropriate alterations both of internal conditions and of such external ones as light, temperature, and humidity. The first-formed tubers are most likely to respond to such changes, while those which appear later resemble more and more the usual egg-shaped or spherical spring tubers. This means that the plant sooner or later adapts to the new conditions and grows again "normally" (using the term according to *present-day* morphogenesis). The atavistic alteration of tubers is, therefore, possible only in the first stages of their individual development, where it is the consequence of some change in the metabolic turnover. It would be very valuable to know more about these metabolic changes and to complement such growth studies with biophysical experiments, in order to establish the interactions among various substances and to be able eventually to induce a particular shape of roots or of leaves. For the moment we have to be satisfied with the conclusion that here also inhibitions lead to the appearance of ancestral shapes.

This conclusion agrees with the concept of comparative morphologists that root tubers are derived from reserve roots. That is the case in Ficaria, where the altered root system does not branch out for a considerable time; instead, the thick, smooth base becomes larger and larger and fills up with reserve substances. The thin end part, branched at first, becomes shorter and shorter, does not develop new branches, and finally rounds off completely (Troll, 1943). In Ficaria all these phylogenetic transitions can be brought about experimentally, so that tubers, which appear later than those which have been transformed into roots, become slightly thinner at the tip but do not branch out; eventually they come to resemble the tuberous roots of *Ranunculus illyricus*, a form undoubtedly more primitive than *Ficaria verna*.

It was for this reason that Troll, when comparing our experimentally obtained atavisms with his own scheme of the development of rootlike tubers, came to the conclusion that the task of experimental morphology is to demonstrate in experimental cultures the value of speculations based on comparative studies.

Basically, then, even this comparative morphologist agrees with the affirmation of Timiriazev, mentioned previously, that experimental morphology should be the basis for the teaching of Evolution. However, Timiriazev went on to point out that through morphological discoveries the mechanism of Evolution should be clarified and the study of Evolution promoted. Also, since we can make use of inhibitions to uncover past evolutionary stages, we might arrive at new evolutionary forms through the use of special stimulations. Actually we have been able to obtain the transformation of normal thin roots into tubers by keeping the plants under long-day conditions in wintertime (Fig. 8c). (This experiment is especially successful with continuous strong electric light.) Under these conditions the secondary roots thicken a great deal through cell division of the innermost layers of the primary cortex, on the outer side of the endodermis, and this gives rise to a reserve parenchyma filled with starch grains.

As in the case of the horse chestnut and other woody plants, as well as in *Ficaria verna*, ancestral forms of leaves can be experimentally induced. This is again brought about by the inhibitions which become increasingly established in the whole plant as the foliate stem grows. These inhibitions reach various levels of effectiveness and the experimental results vary accordingly. If we cut small vegetative stems late in the season, then the buds remaining on the base of the stem do not develop but form additional normal scales, followed by primordia of typical kidney-shaped leaves with long petioles. In underwater cultures the inhibitions decrease to such an extent that after the first-formed scales appear there develop elongated parallel-veined leaves (Fig. 8e) which resemble those in the narrow-leaved species of buttercups like *Ranunculus flammula*. In this respect, *Ficaria verna* represents phylogenetically a more recent type than *Ranunculus flammula* or *R. illyricus*. Systematicians who separate the crowfoots from the buttercups (Ranunculus), indeed, place Ranunculus before Ficaria in classification.

The similarity of form is probably not merely an external one, for this would not allow us to make any phylogenetic conclusions, since homoplasia is not necessarily a sign of homology. Rather, the conditions in which Ficaria forms elongated parallel-veined leaves instead of the usual kidney-shaped ones indicate that this change is a reappearance of leaves common to ancestors of all the families included in the group Ranales. This group is, of course, at the beginning of the classification of the dicotyledonous plants and is also probably the ancestor of all the monocotyledons, in which elongate leaf shape prevails. Some paleontologists actually assign the same origin to two large groups of Angiosperms in the polyphyletic development of the plant kingdom (Němec, 1956). All these atavistic changes stress the importance of integration in plants, since here again we disturb the integrated "whole" either through external influences or by blocking normal internal correlations. Plants respond to these interventions only on certain particular sites, which are the sites of the greatest developmental activity.

Seeds are probably most suitable for these experiments, because their embryos will grow at almost any time; in trees or even in the above-mentioned herbaceous plants the annual rhythmicity interferes and confines our experiments to certain periods of the year. Besides, seeds do not exhibit such strong inhibitions because, as the excellent Soviet morphologist Krenke (1950) puts it, they represent structures with the highest "life potential."

The usual example of this, as noted above, is the vetch, which reacts in much the same way as some woody plants which have been operated on in early spring. Following the cotyledons and two scales there come two evenly pinnate leaves, and only then are larger, entire stipules without leaflets formed. There is no transition between these two forms of leaves. One can also observe in the pea, instead of evenly pinnate leaves with tendrils, which have been fully formed already in the seed, trifoliate leaves, similar to those of clover (*Trifolium*). The number of plants that form such leaves depends on the cultural conditions from the beginning of germination.

Several varieties of garden pea, with both yellow or green seeds, have been studied. In a variety called "Přebohatý" (very rich), no trifoliate leaves were formed in an ordinary culture, if the seeds were first soaked in tap water for 12 to 24 hours. But if the seeds had been first soaked in a solution of the hydrazide of maleic acid (MH), trifoliate leaves are formed with great regularity, on almost all experimental plants. The incidence depended on the concentration of the solution (see Fig. 9b).

For example, from seeds soaked in 0.1 percent solution of MH, 11.8 percent of the resulting plants had trifoliate leaves; a 0.05 percent solution gave 21 percent. In another pea variety, the early blooming "Olomutz," trifoliate leaves were much more frequent in the normal condition, so that a 0.10 percent solution of MH gave 27.5 percent, a 0.05 percent solution gave 28 percent, and water 21 percent.

The development of trifoliate leaves depends also on the

Fig. 9. Garden pea (*Pisum sativum*): (*a*) plant from seed soaked in a solution of maleic hydrazide, showing shortened epicotyl and swollen root tip; (*b*) growing individual at a later time, with trifoliate leaves and replacement root directly under the cotyledons; (*c*) seeds soaked in an auxin solution prior to maleic hydrazide, with main root growing and branching; roots also develop from the cotyledons. Corn (*Zea Mays*): (*d*) in short winter days, terminal spike is formed with female flowers near the base of the spike and male flowers above; (*e*) stronger apical cob feminized by short day; (*f*) Blaringhem obtained similarly feminized cobs following amputation of the stems.

amount of reserve food in the seed, for in the same variety of pea this particular form of leaves appeared in 30.4 percent of plants if we kept both cotyledons, but in only 12 percent if one half to two thirds of the cotyledons were removed. Similarly, of plants grown in Knop's nutrient solution 20 percent had trifoliate leaves, while those in distilled water had only 15.2 percent. In outdoor cultures of this variety, without preliminary soaking of seeds and using garden soil, only 4.5 percent of the plants had trifoliate leaves. With seeds given 3 to 4 days of soaking in water to become fully swollen, there were no trifoliate leaves at all. On the other hand if the seeds were kept for 4 days in *running* water, the number of trifoliate leaved plants rose to 46.8 percent. It is of interest to note the location of these leaves on the plant. They are usually on the fifth node (not counting the basal node carrying the cotyledons), because the first two nodes bear evenly pinnate leaves with tendrils which were already preformed in the seeds. More rarely do they appear on the fourth node if their primordia in the seed have not been completely differentiated at the time of germination. The rarest case, however, is to find these leaves either on the third, sixth, or seventh node.

In a very dilute MH solution (0.005 percent), 33 percent of plants of a variety that normally does not form trifoliate leaves at all developed such leaves regularly on the seventh node. Evidently, such a weak solution of maleic hydrazide does not affect leaf primordia that are already well formed, but only the youngest primordia in the apical meristem. If plants that have undergone MH treatment are decapitated at a particular level, axillary branches grow out; these bear trifoliate leaves much lower down, usually on the second node. Thus the primordia of these axillary buds are embryonically much less advanced than the main bud or plumule. The number of trifoliate leaves on such plants is usually very limited, generally only one or two, after which regular evenly pinnate leaves with tendrils are formed. Here again we can speak of phylogenetic recapitulation, obviously linked to inhibitions caused in the plant by the maleic hydrazide. Indeed, this substance is commonly used to inhibit growth, for example, in reserve organs of vegetables, like potatoes, onions, carrots, et cetera. It is applied not when they start to germinate in the relatively warm storage room but 1 or 2 weeks before harvest, by sprinkling the solution on the leaves in order to increase natural inhibitions present at the end of the vegetative period. According to Šebánek (1957) one can check the unwanted aftergrowth of rye (*Secale cereale*) that occurs usually on harvested fields in rainy weather by spraying the field with MH about 10 days before harvest.

The example of the pea plant treated with MH thus shows clearly that ancestral leaf shapes are actually caused by an increase in inhibitions, either natural or artificial. However, even the pea plant soon adapts to them, so that, after one or two atavistic leaves, normal ones follow immediately. Since the trifoliate leaf of the pea plant, belonging to the group Vicieae, resembles that of clover, belonging to the group Trifolieae, the results of these experiments agree with the classification of these groups in the family Fabaceae, in which Trifolieae precede the Vicieae.[2]

On the other hand, among the monocots, which are most advanced in the evolutionary scale, a good example of phylogenetic recapitulation is shown by flowering in maize (Zea Mays). Even under normal cultural conditions this plant shows many differences from the typical flower structure, for it has a male spike on the top and female cobs borne laterally lower down. It is typically more or less monoeicious, and only rarely do bisexual flowers appear (Slováková, 1957). If, however, we grow corn on a regime of short days, either in

[2] This may be supported by Kloz's electrophoretic analysis of proteins, according to which both Vicieae and Trifolieae contain vicilin, while this could not be detected in the species of Phaseoleae and Genisteae which he examined.

a greenhouse during the short winter days, or during the summer, limiting the light periods to 9 hours by shading, then it very regularly forms female cobs at the top and only very few and underdeveloped male spikelets (Fig. 9d, e). Schaffner (1930) noticed that short days tend to cause feminization, in spite of the supposed influence of sex chromosomes. Sometimes two or three female cobs appear close together at the tip of the main stem, behavior which brings out the resemblance between maize and sesame grass (*Tripsacum dactyloides*), sometimes cultivated for seasoning, but growing wild also. According to this, maize is closer to Tripsacum than to teosinte (*Euchlaena mexicana*), which is usually thought of as the ancestor of maize. (It is no longer found in the wild.)

The flower parts obtained by exposure to short photoperiods actually resemble very closely the oldest known remains of corn excavated in 1948 at 2.5 m depth in the Bat Cave in New Mexico. These specimens had small cobs, only 3.5 cm long, with a few rows of grains covered with seed envelopes. The grains formed in short photoperiods are spherical, covered at least halfway by seed envelopes and supported by large glumes on a zigzag cob with usually only two rows of grain. At the top there is an underdeveloped male tassel which falls off very early.

Similar recapitulation can be brought about by operating on the plant, as in the very interesting experiments of the French biologist Blaringhem (1911), who made an extensive study of the changes caused in corn as a result of injury. He would cut off plants near the soil at different times during the summer months and follow the formation of flowers on the replacement shoots; these shoots grew more and more slowly the later in the summer the operation was performed, because with the development of the whole plant inhibitory forces were increasing. Only if the recapitulation was performed in late July or early August did the plant produce new shoots with female cobs at the top (Fig. 9f). With increasing inhibition, chemical balance in the plant is therefore altered in such a way as to give the same sexual change as with short photoperiods.

This behavior of corn demonstrates the significance of plant integration, governed by a strong root system, for the discovery of ancestral forms (for these feminized forms must be regarded as such). As in all the other cases, the root system is here again the main prerequisite for this inhibitory recapitulation.

These inhibitions cannot be compared with those occurring when the root system is not sufficiently developed, resulting in a shortage of its products and of absorbed minerals. Under these conditions growth is very limited and the weak plants form only male spikes, a condition that is the opposite of the ancestral feminization. Goebel mentioned such weakened plants when listing examples of correlation between masculinization and decrease in food content. The inhibitions leading to phylogenetic recapitulation, which have been described, require a completely different set of conditions in order to produce primitive ancestral forms.

CHAPTER III

Correlations, the Fundamental Basis of Integration

Correlations of growth, that is, experimentally determined interactions between various parts of a plant, increase steadily as the plant develops and as new organs are formed. Interacting factors govern the appearance and development of these new organs, and they in turn form new factors influencing other parts of the plant. Thus the network of correlations becomes more and more complicated as each new part of the plant, at a specific stage of its development, becomes bound into the whole.

There is no need to consider this "whole" as metaphysical. Any positive experiment which solves a problem of correlation is more valuable than pure speculation. In such a study it is reasonable to start with the simplest cases and proceed to more complicated ones. From the germinating seed or the developing winter bud we can pass to fully grown forms, and draw conclusions about the continuous chain of correlations, the result of which is the harmonious development of the plant.

Even in the very youngest, experimentally approachable primordia a whole series of correlations is at work, influencing the differentiation of various parts of shoots or buds. In the bud the future plant is indeed more fully developed than it is in a seed, for embryonic development often continues in the seed outside of the mother plant and later during germination. For example, at the time when the root of the pea seedling is breaking into the surface of the ground, the petioles of the cotyledons are elongating and the foundation of the plumule is being laid. Here we can observe very easily the correlations between the various parts of the embryo, for these are closely linked together in one whole and governed by the cotyledons, which contain righ reserves of nutritive materials and numerous active agents.

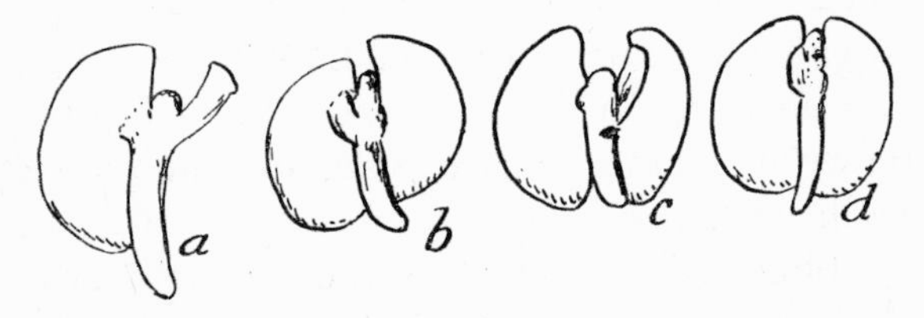

Fig. 10. Garden pea (*Pisum sativum*), operated immediately after the seeds had swollen: (*a*) petiole of the amputated cotyledon elongates much more than the one opposite; (*b*) petiole at right did not grow because the sectioned cotyledon was spread with auxin paste; (*c*) cotyledonary petiole grows more on the side of a transverse cut through the radicle; (*d*) petiole inhibited on the same side as a cut through the plumule.

An example of these correlations is given by the growth of the cotyledonary petioles. In the pea these are opposite to one another and are essentially identical, unless one cotyledon is much enlarged at the expense of the other. (Such *anisocotyledonic* pea seeds are very rare.) Let us first consider the inhibitory influence of the cotyledons, which are rich in nutritive and specific substances and hence represent the most active part of the expanded seed. In the first period of germination the cotyledons actually retard the growth of other organs. If from a seed that has just expanded after 12 to 24 hours of soaking in water, one of the cotyledons is cut off, the petiole on this side grows much more than the one on the side where the cotyledon remains (Fig. 10a). Even after the amputation of half of one cotyledon there is a noticeable difference the next day, for beneath this half the petiole grows more than under the intact cotyledon. This inhibition is linked to the influence of an endogenous auxin, for if we cut both cotyledons in half and spread lanolin paste containing synthetic auxin (indole-3-acetic acid or IAA) on one of the halves and leave the other half bare or covered with water paste, then the petiole on the side treated with auxin grows much less or not at all (Fig. 10b). This is dramatically demonstrated by the fact that the stump without IAA curves downward and the two cut surfaces form with each other an angle which increases with the concentration of IAA in the paste. However, if the auxin paste is very concentrated it now causes inhibition even on the opposite side, so that both petioles grow very little.

The inhibitory effect of cotyledons on growth undoubtedly has great biological significance. Actually it has been observed many times that the reduction of the cotyledonary content speeds up the growth of the embryo at first. Where there are several cotyledons, as in embryonic plants of coniferous trees, the same holds for the removal of some of them. This inhibitory influence of cotyledons is not, however, limited to the cotyledonary petioles, for the radicle also reacts to their inhibitory influence. It grows *more* on the side where the cotyledon has been removed or sectioned. In a vertically growing plant this results in a concave curvature toward the side where the cotyledon remains. If we again cut away half of each cotyledon and apply IAA to one of them, the radicle will grow less on this side. This is surprising since auxin actually favors the development of embryonic roots, and it is frequently used along with other agents ror the artificial formation of roots on a cutting.[1] Here it acts as an inhibitor because elongation

[1] Actually it is not particularly surprising, since, although auxin promotes the formation of roots, it is well known that it inhibits their elongation. Indeed,

is just starting. Only later, when lateral roots form, they will do so first (and grow more strongly) on the side where the cotyledon has been left intact or treated with auxin. Conversely, the radicle of an expanded seed has an inhibitory effect on the cotyledonary petioles; — if we make a transverse cut through one side of the radicle, then the cotyledonary petiole will grow more on that side than on the other (Fig. 10c). Thus the radicle has the power of regulating growth. It slows it down at first, instead of promoting it as does the root system later. Therefore, between the cotyledons and the radicle there are correlations which have probably been established in the seed.

Finally, the embryo contains the plumule, which is at first inhibited in its growth both by the cotyledons and by the radicle. For if we amputate the radicle directly after the seed has expanded, the growth of the plumule is accelerated, that is, its behavior is similar to that of the cotyledonary petioles. On the other hand, the plumule itself promotes the growth of cotyledonary petioles, for if we make a transverse cut on one side the petiole on that side grows less (Fig. 10d).

The primordia of the plant's basic organs, the axis and the root, thus show a remarkable difference in behavior with regard to the correlations which act during embryonic development in the seed. The first to develop are the cotyledons, which then inhibit the formation and growth of other parts of the embryo, especially the radicle. The plumule is undoubtedly less inhibited, because it develops scales which lower the cotyledonary inhibitions and allow the embryonic growth of new stem primordia.

We would have to be thoroughly acquainted with these embryonic correlations to be able to act artificially upon any further development of the plant, for example, by applying

---

if the whole stem of a pea seedling is cut away and IAA applied to the resulting cut surface, it inhibits the elongation of the root and promotes the outgrowth of lateral roots on it. (Ed.) See K. V. Thimann, *Am. J. Botany* 23:561–569, 1936.

stimulating substances. First, we would have to reproduce as best we could the natural conditions. Inhibitions are at work which allow the activity of only certain members of the enzymatic system favoring further growth and development. Hence it is possible, through the use of auxins or similar substances, to check the initial growth somewhat, in other words, to increase the initial inhibitions. In the course of further development the retardation of growth is counteracted, and in many cases replaced, by a more active growth or even by greater productivity of the plant. According to the results of Šebánek (1956) and other workers, we might most successfully use synthetic stimulators to interfere with existing interactions between the plumule and the radicle, or perhaps even between the stem and the root system. This is especially true because root and tuber vegetables show, after appropriate stimulation of their seeds, a richer production of underground organs, while aerial parts are less affected. This can be considered as a "correlative stimulation," because its effect can be explained by favorable interactions between the underground and the aerial organs.

The same is true in the case of the stimulation of sprouting in potato plants (*Solanum tuberosum*) with unsaturated hydrocarbons, especially ethylene. If this gas is added in very small amounts to the atmosphere in which the tubers are kept for 24 hours, growth of the buds is at first slowed down but later stimulated, and in many varieties the productivity is increased. The same stimulatory effect can be obtained with a mixture of air and 7.5–30 percent of illuminating gas.

The potato is a common experimental object for morphologists. As long ago as 1908, Goebel described a short thick apical bud growing on an intact tuber with numerous strong roots, as contrasted with a thin, long shoot growing from an isolated bud, covered with a few simple leaves arranged vertically one above the other. A lack of reserve material shows also in the poor development of roots on such shoots growing

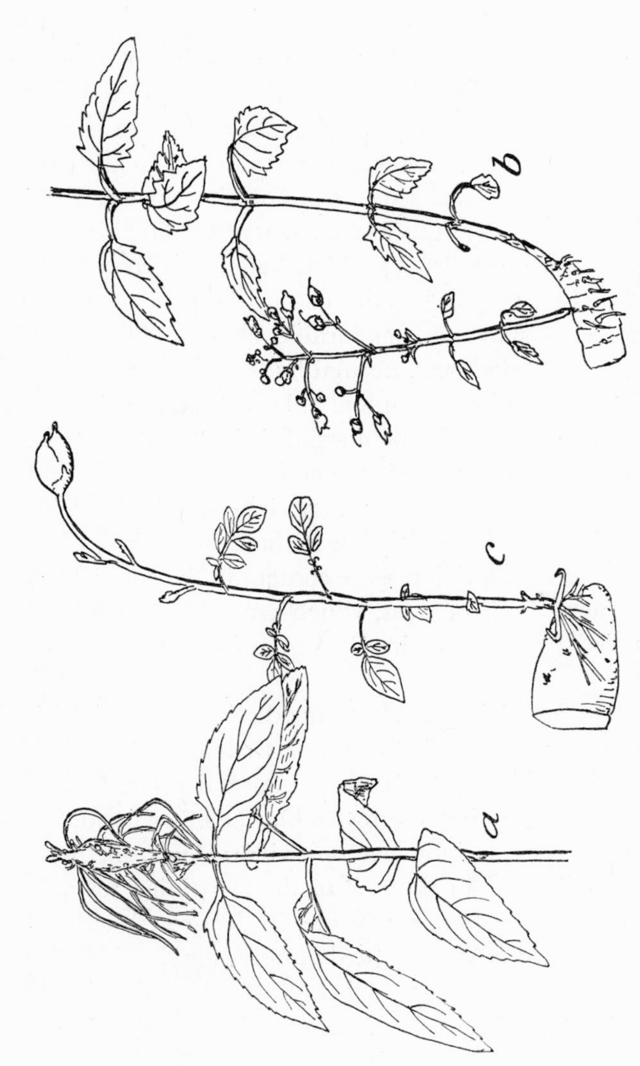

Fig. 11. Figwort (*Scrophularia nodosa*): (*a*) immersed in water, top of stem becomes transformed into a tuber, with roots and stunted buds; (*b*) if the buds are excised from the stem, prior to flowering, an inflorescence develops from the tuber. Potato (*Solanum tuberosum*): (*c*) when grown in the dark, the apical buds formed a runner with a terminal tuber.

from isolated buds. The amount of reserve substances therefore governs the balance of correlations between aerial and underground organs of the plant.

Vöchting (1908), by preventing the formation of underground tubers, brought about the transformation of axillary buds into lateral tubers, covered by inhibited foliage leaves. Sometimes this even occurred near the stem apex. In our own experiments we have succeeded in transforming the *apical bud* of the potato into a tuber; this had its own apical bud, formed from irregularly pinnate leaves which had been irreversibly differentiated (Fig. 11c) before this change in form occurred. *Scrophularia nodosa* shows a comparable change (Fig. 11a). After a short rest period new, normal bud scales were formed on the apical buds of the terminal tubers. These scales are necessary for the formation of normal foliage leaves above them. Thus the same interactions occur between scales and foliage leaves as between reserve materials, which cause inhibitions, and the scales which lower these inhibitions. The embryos of seeds with a rich reserve content in the cotyledons, or the developing buds of trees, which are subject to inhibitions produced by the growing leaves, offer additional examples.

Woody plants which branch sympodially exemplify this phenomenon very markedly. Leaves on their branches cause the degeneration of apical buds, even though these buds were dominant at the beginning of growth. On the other hand, the leaves have a favorable influence on axillary buds, for if we remove a leaf the bud in the axil of the remaining leaf enlarges. This suggests a certain passivity in the growth of buds during their embryonic phase. Furthermore, the floral nature of a bud is influenced not only by the leaf subtending it but also by other parts of the plant, such as the root system. In the next differentiation period the root system also determines the correlative dominance of the apical winter buds. If the buds at the topmost node are cut off but the protective leaves

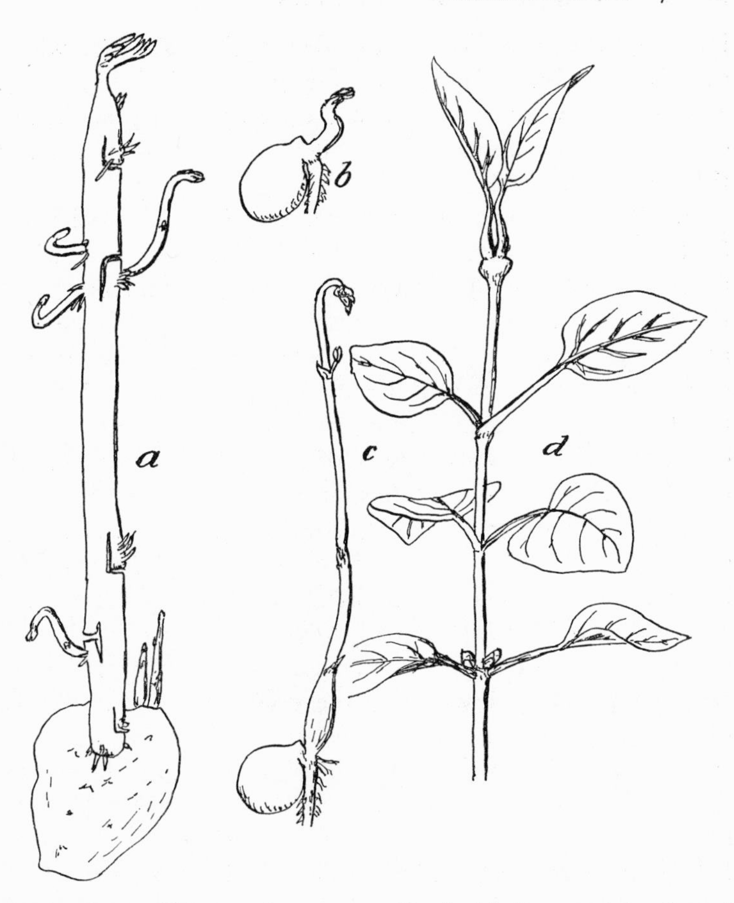

Fig. 12. Potato (*Solanum tuberosum*, var. *Fram*): (*a*) downward bending of small lateral branches below the incisions, showing the stimulatory effect of plastic substances coming from the mother tuber and the roots; upward bending of the small branches in direct communication with the apical bud, showing the inhibitory effect of the apical bud. Garden pea (*Pisum sativum*): (*b* and *c*) after treatment of the plumule on the reverse side, the sharp curvature caused by positive geotropism became reversed (cf. p. 84). Lilac (*Syringa vulgaris*): (*d*) after excision of the buds from the upper nodes, the topmost node, together with the base of the top leaves, enlarges and the leaves stand up vertically, under the influence of the roots (cf. p. 50).

left on, then these leaves, together with the node, widen greatly at the base of the petiole and grow upright (Fig. 12d). If the axillary buds are left on, then the leaves stand out obliquely, as do also the branches that grow from that site the next year.

The lilac bush offers a good demonstration of this behavior, for the opposite arrangement of the leaves permits comparison between the development of buds situated on the two sides of the same node. The topmost buds can grow out to a considerable extent if we remove the leaves from the upper end; they can even form branches in the absence of supporting leaves if the plant is strong enough. On the other hand, buds situated on lower nodes or in the axils of foliage leaves remain undeveloped. This behavior offers fairly clear evidence that integration in plants occurs in buds as early as the embryonic stages. The size of the buds decreases regularly from top to bottom, owing perhaps to the influence of the root system.

Whether a lilac bud is to be floral or vegetative is predetermined to such an extent that it cannot be reversed. Indeed, for leaf primordia in the great majority of plants, qualitative reversion is impossible, probably owing to strong inhibitions in the lower part of the plant. On the other hand, floral differentiation can be induced in undifferentiated buds, such as those found on tubers of *Scrophularia nodosa*, or on the base of the stems of flax. The procedure is to cut out *all* the bud primordia on the stem at the beginning of the formation of the normal, apical flower (Fig. 11b). Then the nutrient material which remained in the leaves, and which would normally have contributed to the formation of the apical flower, can now be used only in the few remaining bud meristems, situated at the base of the plant. As a result, even purely vegetative buds, located where flowers normally never appear, can be changed into flower buds. The stimulus to flower production can thus find a response in any part of the plant if the correlations are turned in that direction.

It is almost certain that buds of hemp seedlings were forced to turn directly into flowers by grafting them into the crown of an adult plant and at the same time removing all bud primordia on that plant, so that only these grafted bud primordia remained for the floral stimulus to act upon. This would again be an example of a balance of correlations, through which floral differentiation would be preferentially stimulated in the plant's own bud primordia, rather than in the grafted immature buds.

When the plant passes from the embryonic to the elongation phase, during which the organ primordia unfold, new correlations arise following the formation of new organs which now interact with one another.

The first of these correlations can be considered an extension of those that governed the formation of the embryo. During that phase we witnessed first a differentiation between the cells forming roots and those forming aerial parts. This is analogous to what occurs in the zygotes (or "eggs") of algae (*Fucus*, for example), where the first cross-wall separates a chlorophyll-containing cell from a nonpigmented cell, a separation which will later give rise to the rhizoids (Fig. 13). It is

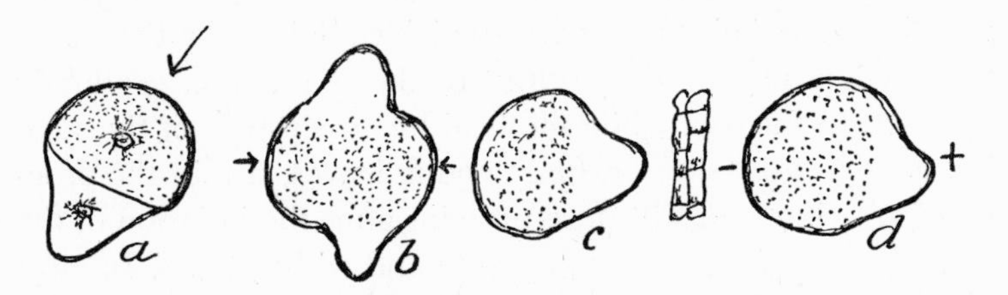

Fig. 13. The brown alga *Fucus*. Diagram of the origin of the polarity in the fertilized egg (zygote): (*a*) light coming from one side determines the orientation of the first division, separating the shaded cell, which will produce rhizoids, from the lighted one, which will form the assimilatory thallus; (*b*) light coming from both sides leads to formation of two primordia of rhizoid and thallus; (*c*) in the vicinity of a piece of thallus or of another egg, a rhizoid arises; (*d*) in an electric field rhizoids develop opposite the positive pole.

evident that interactions between the root and the stem system are particularly important for the integration problem, since these two systems are defined so early in the development of the plant. The same is true of the zygotes of many animals, which differentiate a cranial and a caudal end even before the first division and thus establish permanently the separation between the head and the rest of the body. No doubt specific metabolic differences are established between the stem and the root, corresponding to the respective functions of these two basic organs. It is the task of physiology to elucidate these metabolic differences and enable us to change the correlations between the root and the stem; it is on these correlations that the productivity of the plant mainly depends.

The root emerges first from the seed. This is probably an expression of its metabolism, for it is absolutely necessary for the future growth of the plant. It is not merely the uptake of water and mineral salts from the soil that are important here but also the formation of special products in the root which enable the stem to grow. The root itself depends, however, on the leaves from which it receives carbohydrates and vitamins. Stems deprived of roots do not stay alive for long. They have continuously to obtain proteins for growth by breaking down proteins from their lower parts, which explains why the leaves turn yellow first at the bottom of the stem, then gradually upwards to the top, while the healthy, green part of the stem becomes steadily shorter. Only young leaves, still in the process of growing, can still store proteins after the isolation of the stem; but finally they also grow old and wither away. In most plants leaves give rise to roots quite easily, but the replacement of buds is much less frequent. By forming roots, the leaves, centers of carbohydrate production, make up for their insufficient amino acid production.

The interactions between leaves and roots are shown even in isolated cultures. A well-rooted leaf, even if it is thin and herblike, can stay alive for years, while it soon dies if left on the plant. There is thus a direct influence of the root on the stem. In the opposite direction, the stem has an important influence on the root; this has been known for a long time, the first experiments having probably been carried out by Knight (1806), well known for the first experiments on the cause of geotropism.

In a very schematic way, the favorable effect of roots can be demonstrated by simple experiments on very young flax plants. If the main bud is cut off above the cotyledons, as well as one of the cotyledons, but the two opposite axillary buds are retained, then the bud in the axil of the remaining cotyledon grows more and inhibits the growth of the opposite bud which is in the axil of the amputated cotyledon (Fig. 14a).

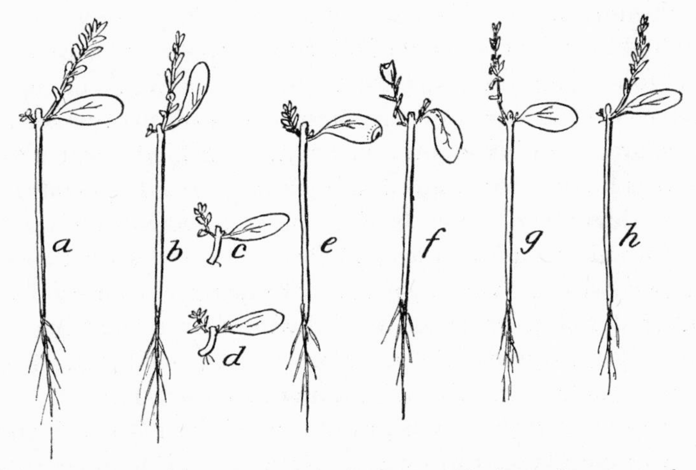

Fig. 14. Flax (*Linum usitatissimum*), seedlings with the main stem and one cotyledon removed: (*a*) in white light the cotyledon *stimulates* the growth of its own axillary bud; (*b*) the same occurs when the stem is treated with auxin paste; (*c*) when the root has been excised the cotyledon *inhibits* the growth of its own axillary bud; (*d*) the same result as *c* occurs when the cotyledon is treated with auxin paste; (*e*) when a drop of auxin paste is placed on the surface of the cotyledon, the bud in the axil is also inhibited; (*f*) the same occurs when the cotyledon is treated with triiodobenzoic acid paste, and a fused, pitcher-shaped leaf results; (*g*) in red light the cotyledon inhibits the growth of its axillary bud (in contrast to *a*); (*h*) gibberellin paste does not alter the stimulatory effect of the cotyledon in *a*.

If, however, the root is also cut off and the plants floated on water, leaving a small segment of hypocotyl, just the contrary occurs; the bud in the axil of the amputated cotyledon grows as a rule much more than the other one (Fig. 14c). Under these circumstances, the cotyledon *inhibits* the growth of its own bud, while in the first case it *favors* it when the root is preserved.

This is certainly an expression of difference in metabolism between differently treated plants. The root of the young flax plant raises the level of nitrogen in the plant and increases the production and incorporation of amino acids into proteins, even when the treated plants have their roots in distilled water so that they cannot take up nitrogen from the environment.

Credit for the discovery of this special root metabolism goes first to the Russian physiologist Kursanov (1952). The results obtained from our little plants grown in distilled water are comparable to his reports on aerial roots of tropical plants such as Ficus, which also contain much more nitrogenous material than any other parts of the plant, although they do not take up nitrogen from the air. Our experiments show that the same cotyledon (which in flax looks like a foliage leaf) will either stimulate or inhibit the growth of its axillary bud depending on whether the roots are present or absent. It is a very delicate correlative inhibition or stimulation.[2]

We see here the first signs of one of the most important growth correlations in plants, namely apical dominance. In this case it has been caused by the cotyledon, which acts differently if the root is present and protein metabolism prevails than when the root is absent and carbohydrate metabolism prevails.

We have found no substitute for the stimulatory action of

[2] Many comparable disturbances of the delicate balance between the buds in the two cotyledonary axils have been described by Champagnat: Removal of some of the young leaves in the bud has the same effect, reversing the sign of the inhibition, as does removal of the roots. (Ed.)

roots. Even if auxin (IAA) is applied to the remaining cotyledon of a rootless plant floating on distilled water, the growth of the bud in the axil of that cotyledon is not increased (Fig. 14d).[3]

It must be emphasized that flax, which has aerial, leafy cotyledons, reacts to auxin very differently from the pea, in which the cotyledons are fat and full of reserve materials. Although in the pea, as already mentioned, a minute amount of auxin confers on the cotyledons a stronger inhibitory influence, in the case of flax even rather concentrated auxin pastes are unable to wipe out the stimulatory effect of the cotyledon (Fig. 14b).

If the flax seedlings are kept in red light, then even with the roots intact the cotyledons can act as inhibitors; for in those conditions, as Timiriazev has shown, mostly carbohydrates are formed (Fig. 14g). Thus apparently a high level of carbohydrates disturbs the action of the cotyledons; in fact in white light we found more carbohydrates in the rootless plants than in those with roots intact. Blue light, on the other hand, stimulates the synthesis of amino acids and proteins; therefore these young plants behave the same way in blue light as in white light: that is, a cotyledon stimulates the growth of its own axillary bud. It is worth mentioning that if both cotyledons are left on the decapitated flax seedlings, but one of them is put in the dark (wrapping it in black paper), the bud in the axil of the lighted cotyledon gains predominance, while the one of the shaded cotyledon is inhibited.

Finally, it is an interesting fact that, even in white light, a plant with roots will show the inhibitory effect of the remaining cotyledon if we apply on its tip a droplet of lanolin paste containing 2-methyl-4-chlorophenoxyacetic acid, which is a well-known herbicide (used under the name of Dikotex or Methoxone in Great Britain for the extermination of weeds in

[3] It would not be expected to be increased, since auxin always inhibits the growth of lateral buds. (Ed.)

fields of various crops, including flax). In low concentration this compound kills only the cells closest to the droplet of paste, but an inhibitory influence spreads from that spot all over the cotyledon, canceling out the action of the native auxin, which stimulates the growth of the bud of flax. The bud of this remaining cotyledon is thus inhibited and the dominance passes to the bud of the amputated cotyledon. This bud, however, eventually feels the effect of the herbicide, which spreads through the whole plant, and as a result it will grow less than it would on control plants. A similar effect can be obtained with triiodobenzoic acid, which in some respects is an antagonist of auxins (Fig. 14f).

Flax seedlings can help us understand many of the regulatory effects observed on other plants, since from them we find that carbohydrates belong to the category of inhibitory substances in plants, while amino acids, and undoubtedly also other nitrogenous products such as nucleic acids, stimulate the development of aerial parts. In the spring, when the leaves of our most common deciduous plants are still not completely developed and photosynthesis is not yet very efficient, although the rest period inhibitions have already disappeared, the intensive development of roots results in an active growth of aerial parts. For this reason, woody plants may show a sudden increase in the number of winter buds, as if these buds no longer inhibited one another. But as soon as the leaves reach a certain size, even though photosynthesis may still not be very efficient, the growth of the shoot slows down. This is the cause of the decrease in the growth of roots, which in many plants rest over the summer months, and it contributes to the fact that the growth of trees in our country stops so suddenly.

The effect of leaves of various ages on roots can be explored by comparing the processes of rooting on cuttings with young growing leaves and on those (taken from the same plant) with fully developed, actively assimilating leaves. It is desirable to select plants of appropriate size and anatomy, as, for example, *Scrophularia nodosa*, which has a square stem that permits us to distinguish the roots formed on each side. On cuttings with one node and one pair of leaves, one of which has been removed, we observe that roots appear in water culture below the remaining leaf, if it is young and still growing, but below the amputated leaf, if the remaining leaf is fully grown and assimilating. We know that intensively growing young leaves which are still forming protoplasm are the site of protein synthesis, while fully grown leaves produce mainly carbohydrates.

This may be related to the fact that of the two opposite buds on shoots isolated from near the growing tip of a stem, the bud in the axil of the remaining leaf grows the more (Fig. 15c), while on cuttings taken from the mature lower part of the stem this bud becomes inhibited by its opposite number, the

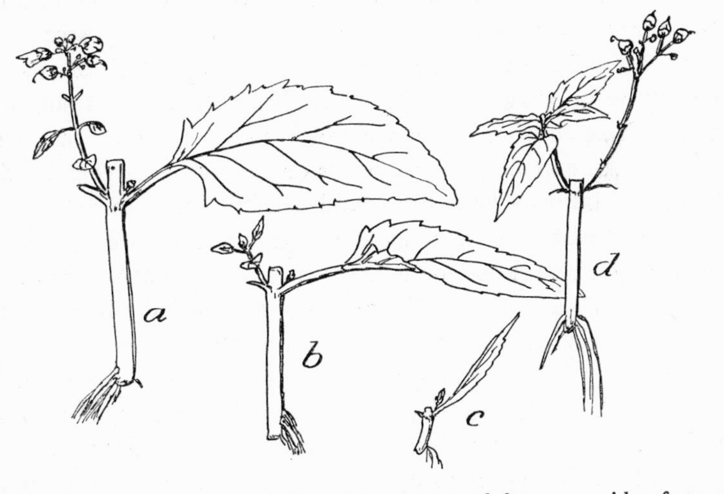

Fig. 15. *Scrophularia nodosa*, leaves of varying age left on one side of stem segments in water culture: (*a*) fully grown leaf inhibits the growth of its axillary bud and the formation of roots below it; (*b*) leaf which is not fully developed inhibits the bud but not the roots; (*c*) very young leaf promotes the growth of its axillary bud and of the roots under it; (*d*) of two shoots on the same node the one bearing leaves elongates much less than the defoliated one, which develops flowers.

one in the axil of the amputated leaf, which grows the more intensively (Fig. 15a). Between these two parts of the stem lies a zone in which the leaves are gradually reaching the end of their growth. Sections from this part show a behavior intermediate between the two previous cases; although more roots develop below the remaining leaf, the bud in the axil of that leaf is inhibited and grows less than the one on the amputated side (Fig. 15b). Thus leaves in the process of maturation stimulate the formation and growth of roots, but inhibit the growth of buds.

Young growing Scrophularia leaves behave therefore in the same way as the cotyledons of young flax seedlings with intact roots (Fig. 14a above). They stimulate not only the growth of future young bud primordia on the stem, but also the growth of the roots, specifically of their youngest parts, which correspond in their amino acid metabolism to the opposite end of the plant bearing the youngest leaves. Between these two opposite poles of the plant lies its whole body, with predominantly fully grown leaves, possessing inhibitory functions. From these considerations we can begin to see why in trees, whose lives extend over hundreds or thousands of years, only a few parts grow at a time, and always during a relatively short interval; thus the resting period is by far the major one, even though photosynthesis and transpiration continue under favorable conditions all through the vegetative period. Discovery of the essential nature of the inhibitions produced by assimilating leaves would have great practical significance.

It is not only photosynthesis and transpiration that are involved, although experiments on Scrophularia might seem to suggest so. If on a section taken from the mature zone of a stem, one of the two leaves is covered with black paper, or if the transpiration in it is reduced by placing it under shallow water, then the bud on the side of the untreated leaf will grow less than the one whose leaf has been covered or is barely transpiring.

The Viennese physiologist Wiesner (1905) explained the inhibitory effect of leaves on buds by saying that the organs that would use up water at the expense of the buds are no longer present and the buds can thus grow. In his day nothing was yet known about the chemical composition of specific inhibitory factors, and correlations were explained by changes in the content of the usual food substances, mainly water and carbohydrates.

Our idea as to the essence of inhibitions was confirmed by experiments on embryonic plants of the garden pea from which we removed the main stem and one cotyledon. Then under favorable natural conditions for pea germination, in the dark or in very dim light, the bud in the axil of the amputated cotyledon grew more than its opposite number, and inhibited the growth of the axillary bud of the remaining cotyledon (Fig. 16a). The latter bud is susceptible to cotyledonary

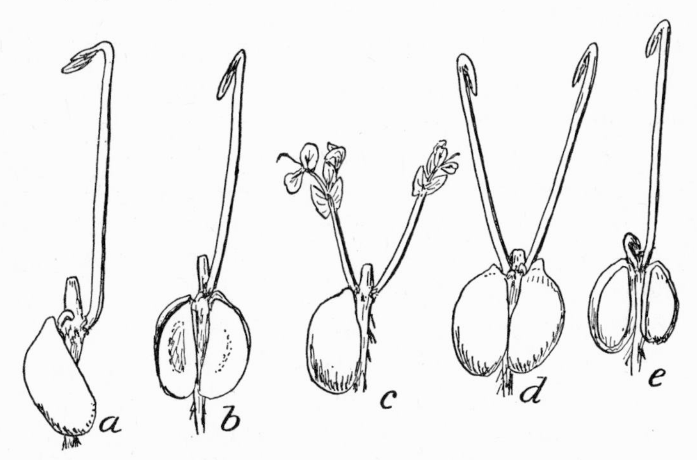

Fig. 16. Garden pea (*Pisum sativum*), decapitated seedlings: (*a*) the shoot in the axil of an amputated cotyledon grows larger than the one opposite; (*b*) when one cotyledon is treated with auxin paste (left), the shoot on that side is inhibited; (*c*) in strong light both shoots grow equally, even after the amputation of one cotyledon; (*d*) when both cotyledons are present the two buds grow equally for a long time; (*e*) when part of each cotyledon is removed, one of the buds soon becomes dominant over the other.

inhibition at a very early stage of its growth and gives way very easily to the dominance of the other bud.[4] This same correlation will apply even if the operated seedlings are kept in running tap water, a fact which disproves Wiesner's idea of correlative inhibition by way of transpiration. Instead, specific inhibitors are involved, and they are not merely the reserve contents of cotyledons (starch and proteins), as one might be led to believe from the inhibitions produced by assimilatory green leaves or by the scales on the tuber of *Circaea intermedia* (Fig. 24c, d). This can be proved by an experiment similar to the petiole test. We take a decapitated pea seedling, cut off both cotyledons identically and apply on one of them a small quantity of auxin paste (Fig. 16b); one gamma, 10^{-6} gm, of this stimulator is sufficient to inhibit the corresponding bud and thus allow dominance by the opposite bud (whose cotyledon was untreated or treated with plain lanolin).

Thus we have experimentally established correlations of two kinds: correlative inhibition by the remaining cotyledon, or by a portion of it plus auxin, and also the dominance of the bud in the axil of the amputated cotyledon. The latter can be compared to the original apical dominance. We would expect the bud to grow better in the axil of the remaining cotyledon, for it is closer to the food supply necessary for its elongation. At first, both of the tiny bud primordia grow at about the same rate, but the difference in size becomes apparent at 1 or 2 millimeters' length. Thus the bud in the axil of the remaining cotyledon must be inhibited by a specific substance produced by that cotyledon. Meanwhile the opposite bud is better able to activate its natural auxin and thus

[4] The ease with which the bud in the cotyledonary axil can be inhibited by small amounts of auxin (whether applied artificially or produced naturally) has engendered an extensive literature (see K. V. Thimann, "Auxins and the inhibition of plant growth," *Biol. Rev. Cambridge Phil. Soc.* 14:314–337, 1939). (Ed.)

to complete the inhibition of the bud in the axil of the remaining cotyledon.

Stimulating substances appear much later during germination. If we keep both cotyledons on embryonic pea plants, the two opposite buds on a decapitated plant do not grow quite equally, but one eventually acquires predominance over the other. The cause of this phenomenon has been clarified by Kořínek (1922), who discovered that if both cotyledons are intact the two opposite buds grow for a long time at the same rate, while if we trim both cotyledons down to one half or one third of their size, one of the buds will immediately dominate over the other one. He designated the ratio of the weights between shoots of various sizes as the measure of "correlative sensitivity" (Fig. 16d and e).

If both buds have enough food supply, the ratio of their weights is roughly equal to one. When the food supply is reduced this number becomes higher and higher as the organic and mineral contents of the cotyledons get lower. In this way the plants can develop at least one bud if only at the expense of the other, which probably demonstrates the main usefulness of these correlations.

We must not see these phenomena as the expression of some supernatural will in nature; on the contrary, these simple experiments constitute a proof of the material basis of correlations. This becomes evident when the production of axillary buds on plants from which *one whole* cotyledon has been removed is compared with those from which *one half* of *both* cotyledons was removed. In both cases the two buds which grow after the amputation of the main stem receive the same quantity of cotyledonary reserve substances; but in the first case the dominance of one bud is established much sooner than in the second, where the two buds fight for the dominance for a long time. Actually the weight of the new shoot formed on plants with one whole cotyledon is greater than that of the shoots formed on a plant with two half-cotyledons. The sensitivity of correlation increases with the decrease in translocated material.

New shoots growing from old stumps of trees show also how in the presence of excess food a great many normal and adventitious primordia grow at first with no control; we can explain such hypertrophies as the result of an altered harmony between individual primordia when food and growth substances are in excess.

These growth interactions are by no means independent of external environmental conditions, for generally all correlations are subject to them. Again the simplest example is given by an embryonic pea plant. We remove the main stem and one cotyledon, but we expose both axillary buds and the remaining cotyledon to very strong light. The two opposite buds will then very often grow at the same rate. The bud in the axil of the remaining cotyledon is not inhibited; on the contrary it often gains priority over the other one, a phenomenon never observed in the dark (Fig. 16c). In fact generally all plants, woody or herbaceous, tend to branch more in strong light, because light greatly decreases the dominance and enables lower lateral buds, which usually remain dormant, to grow out. This is of great significance in field or forest work: the forester with a thicker plantation achieves trunks without knots while the agriculturist with an appropriate row width increases both the quality and quantity of his plant's production.

There is an analogy between our experiment with pea plants bearing only one cotyledon and kept in strong light and the case of isolated fir trees, on mountains, which possess big branches even at the base of the trunk. These branches sometimes produce roots at their extremities and then grow upwards, so that whole families of fir trees can arise. Just as the bud in the axil of the amputated cotyledon loses its inhibitory influence, so the tip of this fir tree loses its apical dominance. The process is further stimulated by the ultraviolet rays in mountain sunlight, which disturb the activation of auxin in the tip and thus decrease its dominance over lateral branches. It is also affected by the preponderance of the root system, noticeable already in the tips of those lower branches which become independent under the influence of their own roots. From all this we see that apical dominance is influenced by many factors and is not, therefore, a kind of sovereign correlation, as it is sometimes thought to be. In a plant everything is related to everything else — nothing governs without being governed. But in individual parts it is often difficult to isolate those interactions which contribute to a harmonious whole; they have to be discovered in models, such as our experimental pea seedlings.

Another phenomenon is brought to light on such young pea seedlings by removing most of the main stem, cutting both cotyledons in half, and applying auxin paste on one of them. At the same time we make a longitudinal cut through the remaining part of the main stem, slide in a piece of glass to separate the two halves, and apply the same amount of auxin to the stump on the side where the cotyledon is untreated. In this way one of the buds is exposed to the inhibitory influence of auxin from the cotyledon and the other to that from the stump of the shoot (Fig. 17c, a, d). We would expect that the axillary bud on the side of the treated stump, under which a tumor with roots soon forms, would be the one to grow less, for Thimann and Skoog (1933) showed on decapitated bean plants (*Vicia faba*) that the cotyledonary buds did not grow when the apical bud was replaced by auxin (Fig. 17b). In our experiment we find, however, that the bud below the treated half of the stump grows more than the other; that is, auxin on the cotyledon inhibits the growth of its own bud, so that the opposite one gains predominance. With the same dose of auxin, then, this substance acts as an inhibitor through the cotyledon more effectively than through the epicotyl.

From this it might be concluded that growth correlations are governed by the cotyledons and that apical dominance is in fact a consequence of their regulatory effect (Fig. 17a).

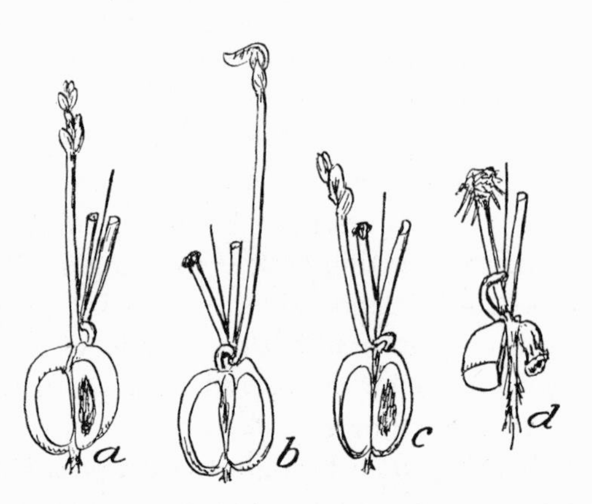

Fig. 17. Garden pea (*Pisum sativum*), decapitated plants, with the stub of the stem split longitudinally: (*a*) when auxin paste is applied on one cotyledon (right), its axillary bud grows less than the one opposite (cf. Fig. 16*a*); (*b*) when auxin paste is applied on one side of the stump (left), the shoot on the opposite side grows the more; (*c*) when auxin paste is applied simultaneously on one cotyledon (right) and on the stump on the opposite side (left), the shoot which grows most is on the side of the treated stump (left); (*d*) same experiment, but using 0.5 percent rather than 0.05 percent auxin paste. The piece of cotyledon treated with this paste becomes emptied; the treated side of the stump forms a large tumor and roots, and the cotyledonary bud grows out on the same side.

If we did not know that the cotyledon has a greater inhibitory effect than the epicotyl we would not be able to understand why the dominance, first established by the cotyledon, can in the course of later development change or be suppressed under similar regulatory influences. It is understandable also that branches of trees, which during the first opening of buds form leaves without any reciprocal inhibition, stop growing later in the summer although growth conditions have become even better. In the same way as shown by cotyledons in the previous model experiments, leaves inhibit not only their own axillary buds but also the apical bud, so that both types of bud will turn into winter buds, or, sometimes, under especially strong inhibitions, the tip will dry out and sympodial branching will follow.

Scrophularia furnishes a good example of the regulation of the growth of shoots. If in the spring the leaf primordia are removed from the apical bud of germinating tubers, then the elongation of the internodes is greatly reduced and axillary buds, which would normally be dormant, grow out instead of leaves. The largest buds grow on those nodes where, if the plant were intact, leaves would also have grown most strongly. It is interesting to note that these axillary shoots simulate leaves even in their opening movements. They do not grow upright; rather, their tips bend in a semicircle to a horizontal position just as leaves in their youngest stages of development turn upwards and only later bend away from the stem by their epinastic growth.

When at any particular node the growth of axillary shoots has become most active, these shoots then inhibit the growth of all other axillary shoots and buds, not only the lower ones but even those above, which will never increase in size even if we keep removing the leaves from the tip. The single main stem thus becomes replaced by two lateral stems growing out from the node where at the time of the operation growth was the most intense.

On an intact plant conditions are not so simple, because leaves, unlike shoots, have a limited growth, so that growth of the leaves at a certain node always stops and is transferred to the node higher up. Up to a certain level on the stem the development of leaves and internodes is continuous, because at the time of the intensive spring growth of the stem the root system becomes stronger and stronger. On the other hand, branches on a leafless stem, which have unlimited growth, inhibit the growth of buds on the nodes above them, so that the growth of stems is finally always limited. In Scrophularia the number of leaf pairs does not exceed eight to ten. Only under conditions which particularly stimulate vegetative growth and decrease leaf inhibition can this otherwise sharply limited plant continue to grow indefinitely. In the dim light and high humidity of the greenhouse, such plants have formed thirty leaf pairs, and growth was still not at an end. Even these leaves were able to inhibit their axillary buds, although they stimulated the growth of new leaves situated above them.

The steady upward shift of the center of growth intensity on the stem can be demonstrated on a very young stem of Scrophularia, sprouting from a tuber, by leaving on the lowest leaves (incidentally still very young) but removing all the other ones. Then, as a rule, growth will at first be most rapid in the axillary buds of the highest remaining leaf pair, but later the maximum will shift to a higher, leafless, node, and it is from this that the longest shoots will grow. The axillary shoots will replace the main stem, which, following repeated defoliation of the tip, will have stopped growing. Just as in this case the lowest remaining leaves transferred the growth of axillary shoots to the next node above them, so also leaves on an intact leafy stem gradually transfer growth activity upwards on the stem. Only if the growth of the tip is completely inhibited does this transference not take place.

The regulatory influence of leaves on the stem at various stages of development can be shown by removing all leaves but one (Fig. 18a), retaining only the very small leaf primordia at the apex. The loss of so many leaves is compensated, in the light, by the growth of lateral buds. If we defoliate in this way a bud on the tuber of Scrophularia, when it begins to unfold, or on a shoot of lilac at the time when it starts to elongate, we find that the buds situated on the side where the one leaf remains now grow more than those opposite; this holds not only for those actually in the axil of that leaf but also for those on the nodes above. At the same time, the axis of the young stem also grows more on the side of the remaining leaf, so that a convex curvature develops. The young growing leaf, therefore, at the beginning of the vegetative period, stimulates the growth both of the main axis and of the buds, perhaps by supplying food materials which are very abundant at that period. The growing apical leaves also show a predominance on the side of the remaining leaf.

If this operation is performed a little later but the leaf retained is still growing, it will have the same favorable effect

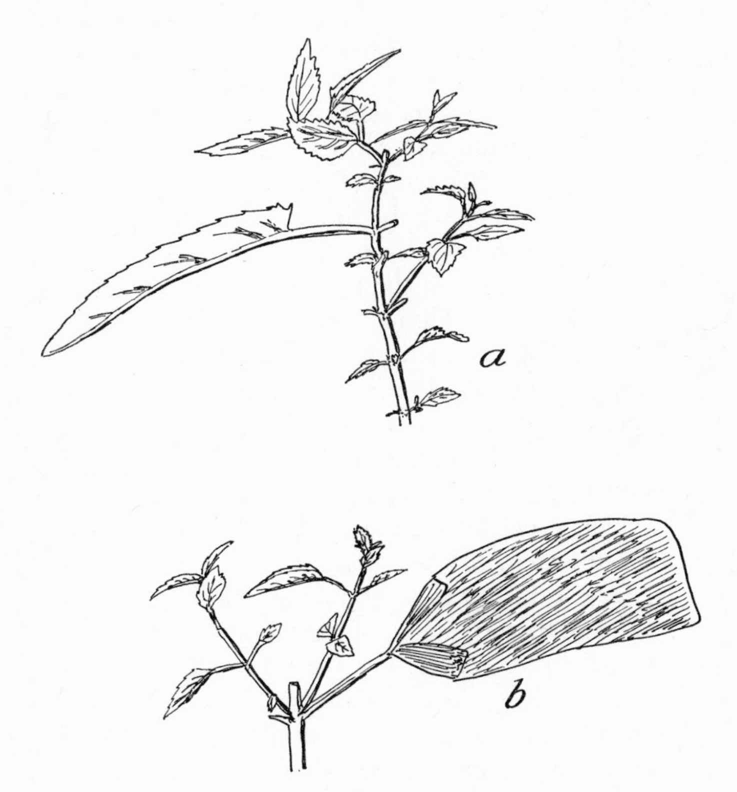

Fig. 18. *Scrophularia nodosa:* (*a*) decapitated young stem, on which only one leaf and no bud has been left; growth is inhibited below the leaf and stimulated above it; (*b*) isolated stem segment with one leaf covered with black paper; both buds grow into shoots which bear, on the second node, an asymmetrical pair of leaves.

on its axillary bud and on buds situated above it on the same side, but the growth of the buds below the leaf will be visibly inhibited. In other words, right under the remaining leaf, the buds grow less because of the inhibition exerted by the shoots above the leaf. The top growing leaves therefore inhibit the growth of buds lower down, just as on an intact plant these leaves inhibit the lower leaves, to produce an upward shift of growth intensity. If the operation is done still later and the retained leaf is now fully grown, its inhibitory action extends not only downward but also upward, so that axillary buds grow more on the side without the leaf, and the apical leaves become bigger on this side.

Such growth phenomena can often be noted in the leaves of lilac bushes growing out of doors if we remove one leaf from the third node below the tip. When the tip continues to develop then the leaves above the lost leaf grow more than those on the other side. If the tip is removed also, then large replacement branches develop on the same side, resulting in anisophylly[5] caused by the inhibitory influences of leaves predominating on one side. In some species of plants anisophylly is a common phenomenon, for example, *Goldfussia anisophylla*, which shows a so-called habitual anisophylly, one leaf on each node being always smaller than the other. However, plants which grow from stem fragments remaining on the root after the stem has been cut off are usually isophyllic, that is, they form leaves of equal size on both sides of the node. The stimulatory action of the roots thus obliterates the differences between leaves. On the other hand, on the topmost nodes of a lilac stem one of the two leaves is very often inhibited in its growth and its bud is only very partially formed. If the shoot carries flowers, then this particular bud remains vegetative while the opposite large bud, n the axil of the normal-sized leaf, forms flowers. Even this case of anisophylly

[5] I.e., two leaves which normally would be a pair becoming unequal in size. (Ed.)

is a correlative phenomenon, since it is caused by a primary increase of inhibitions in the plant, which then leads to a difference in the size of the two buds on the same node.

Generally these unequal buds are found on one of the middle nodes of the lilac stem, between the upper nodes bearing large buds and the lower ones which have only small dormant buds. It is notable, however, that if this node is isolated together with the dormant part of the shoot, then the smaller bud is the first to unfold while the larger one remains at rest[6] (Fig. 24e).

The phenomenon of anisophylly can easily be brought about on the stem of Scrophularia if in the spring, before the apical bud of the tuber starts to unfold, we remove all the roots on one side and retain them on the other, or if we remove all the roots and then cut off one half of the tuber. Either of these operations will decrease the usual spring inhibitions on one side (Fig. 19c), so that above the remaining half of the tuber the leaves grow to a smaller size, while in the first case the leaves grow more above the side with the roots than above the other side. These influences, however, only reach to about the middle part of the stem; above this the differences between leaves disappear and are even reversed, because the larger (middle) leaves now inhibit the growth of the upper leaves more than do the lower (smaller) leaves. In this way, somewhere halfway up the stem a reversion occurs, because the stimulated leaves reach a larger size and thereafter inhibit the leaves situated above them.

If all the leaves are removed from a sprouting apical bud of a tuber, then the very small leaves that have been retained reach an extremely large size and their total number increases. For this we have to keep removing the buds on the defoliated lower part, since otherwise they would replace and inhibit the apical growth of the stem. The lower defoliated part has

[6] Evidently, therefore, the large one does not exert an inhibitory influence. (Ed.)

short but very thick internodes while the upper part has longer and thinner ones.

Even young lilac shoots can be forced to unfold pairs of leaves which would usually have been inhibited, by simply removing the leaves from lower nodes early enough. When this is done on a shoot that, because of its position on the shrub, would normally form flowers, it will produce instead

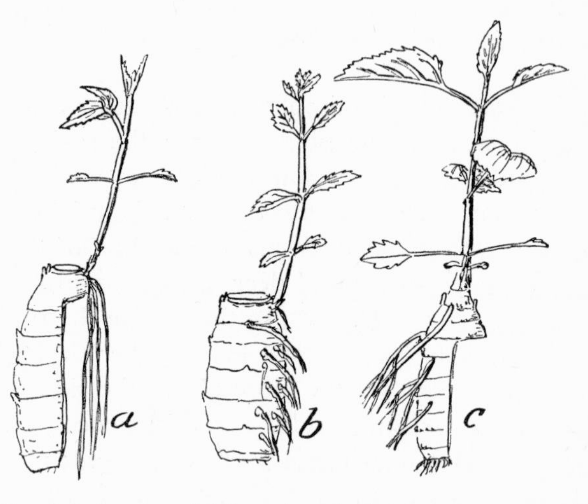

Fig. 19. *Scrophularia nodosa:* (*a*) decapitated tuber, cut out on one side, forms shoot and roots on that side; (*b*) the shoot elongates most on the side where the roots have been left on; (*c*) the leaves grow largest on the side where the roots and the body of the tuber have been left on.

additional pairs of leaves. Those leaves, that on an intact plant would have produced flowers from their axillary buds, retain their vegetative nature; it is only the next set of leaves, which develop as a consequence of defoliation, that produce in their axils primordia of winter flower buds. This operation is successful only if performed before the differentiation of flower primordia in the buds has been completed, that is, before the middle of April. Later defoliation results in the elongation of the differentiated buds, which form, instead of bracts, large leaves, having flower primordia in their axils, but no axillary winter buds for the next vegetative period.

Very strong inhibitions stop apical growth completely. This can be observed in Scrophularia when grown in dry places, especially in the shadow of a deciduous forest; the stem is then composed of long internodes of average thickness and relatively few large leaves. The largest leaves generally develop close below the tip, and then inhibit the growth of new leaf pairs, which will be thickly crowded and separated by short internodes. In this apical rosette of leaves the larger specimens will show flower primordia, but these do not develop. Very probably the large leaves inhibit the growth not only of these buds but also of the roots. It is possible, however, to induce the development of flowers, either by removing these large leaves or by cutting off the stem and placing it in water so that roots develop at the base. Both of these devices decrease the inhibitions so that the tip elongates and flowers unfold.

In nature this change occurs after prolonged rains in late summer or in the fall. The resulting sudden supply of mineral substances decreases inhibitions, so that not only the apical bud but also the upper axillary buds develop. Gibberellin paste applied to the inhibited tip will make it grow also, and if flower primordia have been formed they will open.

Circaea intermedia shows comparable changes. In this case vegetative stems which have been inhibited in their growth by lower leaves will suddenly elongate after application of gibberellin. The resulting new shoots look very much like runners, but they are vertical and bear somewhat larger leaves. Such inhibited vegetative stems do not produce flowers under the influence of the gibberellin, nor do the axillary buds on isolated sections taken from the basal part of the stem. While control segments develop normal horizontal runners, which turn down into the substrate at their extremities and produce tubers, gibberellin-treated buds produce similar runnerlike sprouts, but they grow vertically upward and carry small leaves. Therefore, we conclude that gibberellin does not cause the plants to flower, but only brings about their excessive elongation and the change in geotropic response in the basal parts (cf. Fig. 28e).

Similar correlations are found in vegetative stems of *Bryophyllum crenatum;* in the autumn these form short internodes at the top and later, during the short winter days, a rather thick leaf-rosette. This rosette in dim light may even produce some roots. The following spring the apical bud of the rosette elongates and forms a new leafy stem. In this way growth can continue on the same axis, showing remarkable differences between the various zones according to the day length under which they developed. When all the leaves inhibit the growth of their axillary buds, then the main axis continues to grow monopodially. Thus two different types of sections along the axis alternate: sections with short internodes and round leaves and sections with long internodes and large oval leaves, depending on whether they were formed in short winter days or long summer days, respectively.

This tropical plant, originally from Madagascar, has retained in our climate the rhythmic development of its axis and leaves and the formation of roots, the growth of which, especially near the nodes of short-day zones, is not hampered even by the relatively low humidity of the laboratory atmosphere. It probably represents an adaptation to the even humidity of the tropical climate — an adaptation which has not disappeared under the new conditions, thus showing how deeply rooted the rhythmicity in such plants must be. The development of roots often occurs on the lower side of horizontally growing stems because of the accumulation of auxin on this side; Went (1930) demonstrated on plants of the Bryophyllum group that the formation of roots depended on specific substances accumulating under the influence of gravity on the lower side of the horizontal axis. Stronger specimens of flowering plants show similar conditions of unlimited growth (except that their growth is sympodial), for they are renewed not only from the basal lateral rosettes but also from the upper part of stems bearing fading blossoms, from which lateral branches arise that produce roots on their lower sides.

Both of these cases of growth and branching are analogous to the normal growth of woody plants. Bud scales of woody plants, representing the beginning of each shoot, correspond to short-day leaves, while the foliage leaves on shoots with long internodes correspond to the oval-shaped leaves formed, along with long internodes, on summer stems of plants of the Bryophyllum group. The same analogy is found between sympodially branching bushes and the floral shoots of this well-known exotic experimental plant.

The initials of axillary buds, as well as those of the leaves, lie in the apical bud above the primary leaf primordia, and at first they can still be changed into leaves if their connections with these leaf primordia are severed by an appropriate cut. From the very beginning, therefore, the buds are dependent on their supporting leaves. This dependence is maintained later, so that the axillary buds correspond to a great extent to the nature of their leaves, and change with them along the length of the stem. The embryonic formation of those buds is also dependent on the leaves, for if one very young pair of leaves is removed or even cut very short, there will be no significant change of internodal length, either on the main axis or in the adjacent leaves. The reason for this lack of effect is that when the leaves are removed so early the buds in their axils are not yet far enough developed to grow out and replace their underlying leaves; instead they remain inhibited on the defoliated node by other, normally growing, parts of the plant. On the preceding node, however, the leaves of the axillary buds enlarge, while the internodes of their stems remain quite short. The plant therefore compensates for the loss of one very young pair of leaves by the growth of a bud, not on the same node but on the node below, a bud which is further along in development. Had the leaf pair been retained it would have inhibited these lower buds (Fig. 20a). There is thus an especially delicate correlation

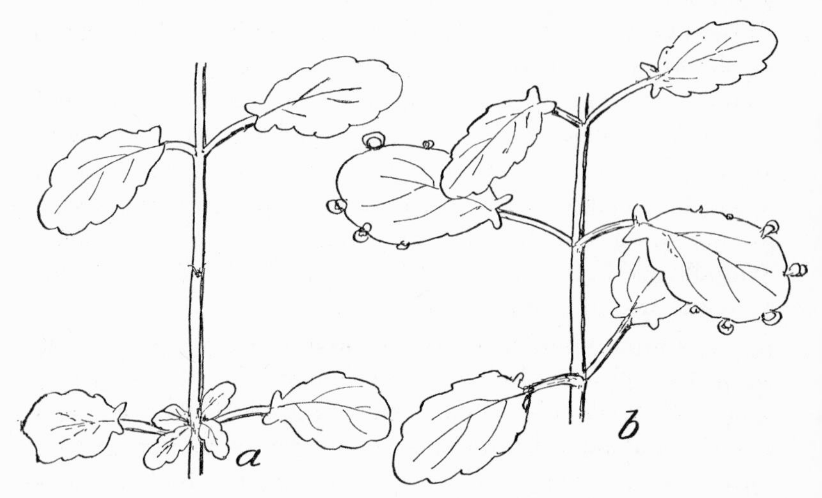

Fig. 20. *Bryophyllum crenatum:* (*a*) removal of the primordia of a leaf pair leads to growth of the buds on the next node below; (*b*) leaves on the middle of the stem (which marks the transition between the lower stimulated zone and the higher inhibited one) bear developing marginal buds.

between individual leaf pairs and their axillary buds, since every new leaf pair inhibits from the very beginning not only the pair of leaves situated next under it but also their axillary buds.

These phenomena indicate a very close relationship between buds and the leaves on neighboring nodes, a relationship which could be demonstrated by the above operation carried out while the leaves, and especially their axillary buds, are still quite young. On an intact stem these relationships between sequential pairs of leaves throw a new light on the metameric development of stems. Plants of the Bryophyllum group are particularly well suited for these experiments, for, unless the specimen is a very large one, their axillary buds are hardly detectable. This indicates the presence of very strong inhibitory influences, which are indeed demonstrated by the marginal buds from which these plants take their genus name and which are regularly distributed around the edges of the leaves, in special notches, as vestiges of embryonic tissue. In the *Bryophyllum crenatum* species these buds develop only if the leaves are isolated from the stem or if they fall off the plant by themselves; they thus ensure rapid vegetative reproduction.

There is an interesting relationship between the formation of these sprouts and the surface of the leaf that produces them. Toward the end of his life, Loeb (1924) tried to prove the so-called Mass Law of growth, which states that the greater the leaf surface the stronger the production of sprouts and roots that grow from it. If we take isolated leaf pairs and in each pair keep one leaf intact, divide the other obliquely into two or three sections, and then grow them in ordinary soil, in sand or even on glass, supplying them with sufficient light (an absolute requirement for this plant), the sprouts arising from the separated sections come to represent twice or even three times as much dry weight altogether as the sprouts regenerated on whole leaves (Fig. 21c). In reality, therefore, the inverse relation is true: the larger the surface of the blade the smaller its productivity.

In large leaves, or larger sectors of leaves, there is a greater accumulation of assimilation products, which slows down the assimilatory action of the chlorophyll apparatus.[7] This is why some reduction of the leaf surface tends to bring about an increase in the plant's productivity — a fact which must certainly have a practical bearing in agriculture. Luxuriously overdeveloped leaves on a crop plant not only shade one another but their products can also decrease photosynthetic

[7] Dr. Dostál goes beyond the facts here, for photosynthesis is not known to be slowed down by the accumulation of products. (Ed.)

activity and therefore the harvest. Long ago Lubimenko (1910) obtained much larger specimens of radishes (Raphanus) by making their leaves smaller.

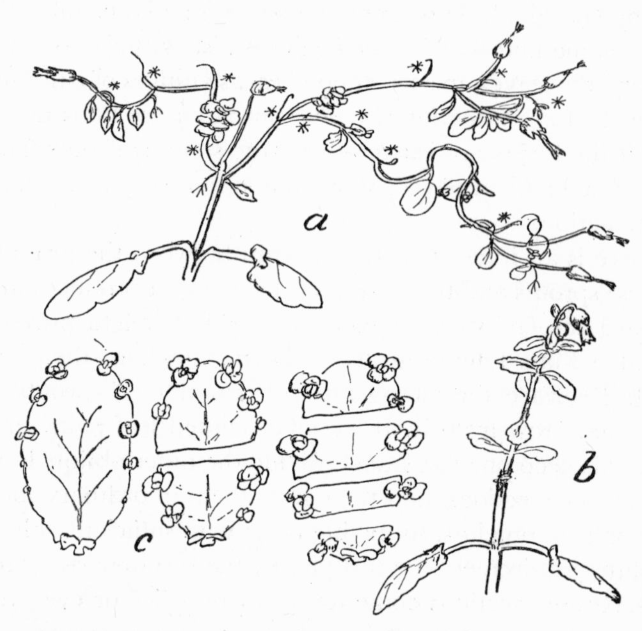

Fig. 21. *Bryophyllum crenatum:* (*a*) the top of a flowering stem, when grafted during a long day on a strong vegetative stock, shows signs of vegetative growth, and unfolds flowers only after it has formed several stunted buds (marked by asterisks); (*b*) a vegetative basal rosette flowers when grafted on a long-day individual, and transferred to short-day; (*c*) sister leaves, one of which is intact and the other divided into two or four parts, show an inverse relationship between the total dry weight of the marginal buds and that of the leaf surfaces which produce them (1 gm dry weight of leaf bears 103 mg dry weight of buds when the leaf is intact, 207 mg when the leaf is halved, and 421 mg when divided in four).

Isolated leaves of Bryophyllum use their assimilation products for the growth of the marginal buds. These do not usually develop on intact plants, as noted above, for they are inhibited both by the apical bud and by the outflow of assimilation products from the leaves to other parts of the plant. However, this behavior changes radically after flowering and maturation of seeds have been completed, for then the leaves continue to assimilate, while the products, which have only been partially used up for the growth of some of the lateral buds, accumulate in the blades. Thus during the winter the marginal buds grow intensively even on intact plants.

The same accumulation of assimilation products occurs in the topmost remaining leaves if the stem above them is cut off. To compensate for the loss of the stem, lateral buds in the topmost axils develop and the marginal buds of their supporting leaves enlarge. It is worth noting that at that time, on just one or two nodes situated at about the middle of a completely intact stem, the marginal buds can become remarkably enlarged, while on both the older and the younger ones they remain invisibly small (Fig. 20b). The enlargement of marginal buds on these middle (usually also the largest) leaves indicates an accumulation or blockage of translocated material, which is retained in those leaves even without their isolation or the decapitation of the stem above them. Such a blockade is also linked to the decrease in leaf size on the main stem, since beyond the leaves bearing enlarged marginal buds the subsequent leaves get gradually smaller as we go up the stem. The flowers, which form at the top when the days get shorter, also contribute to that effect.

The marginal buds start to appear on the leaves at about the time of the autumn equinox, when the flower primordia are formed. This is linked with the appearance of a definite, though rather weak, inhibition occurring at the point of transition between long and short days. Such changes are obviously related to the normal rhythm of these tropical plants, for if the plants are maintained in long days with artificial light through the fall and the winter, the stems will continue to produce long internodes and the leaves will be large egg-shaped blades. In other plants which do not form marginal buds a similar reversion occurs at the level of the largest leaves, approximately in the middle of the stem. The stimulatory effect of the root system, which tends to produce larger and larger leaves toward the top of the stem, is here again counterbalanced by the inhibitory influences coming from the highly developed foliage apparatus, so that toward the top of the stem (whether floral or vegetative) the leaves become smaller.

As already mentioned, this inhibitory influence of the middle or largest leaves is apparent in plants of the Bryophyllum group even after flowering is over, because, although the marginal buds then grow, the lateral buds in the middle part of the stem are inhibited by the largest leaves. At the base of the stem, which is generally leafless by that time, and near the top if the leaves have been preserved, there appear rosettes of the round fleshy leaves which are typical of short-day winter conditions. Only on the largest leaves in the middle of the stem is growth still possible when the leaf's life has nearly ended.

These regulatory conditions in the stem can be influenced by external factors, especially light and humidity. For example, Scrophularia, cultured in weak light and with high humidity of the soil and atmosphere, behaves like other vascular and nonvascular plants well known to morphologists for their continuous vegetative growth under these conditions. *Scrophularia nodosa* forms, one after the other, a great number of leaves, which inhibit the growth of their axillary buds, while the apical bud continues to grow. As many as thirty or forty pairs of leaves will be formed before further formation of leaves or tubers comes to an end. If such plants are transferred to a stronger light, then the apex usually stops growing and from the middle zone of the stem grows a lateral branch which forms flowers. Thus the middle zone of these stems, grown under conditions which do not elicit much inhibition, is the most favorable place for growth to begin again. Similar behavior is seen in strong specimens of woody plants (for example, in the lilac or the privet, *Ligustrum ovalifolium*) which branch out most frequently at the middle level, while in both the lower and the upper parts of the tree the lateral buds remain dormant.

Finally it can be observed, especially on weak specimens of Scrophularia grown with the base of the mother tuber and the roots exposed to light, that when the activity of the roots is thus slowed down the apical growth of the main stem is inhibited. The only region still capable of growth is then the base of the stem, from which a new stem (with a shape like that of the mother stem) grows out, thus bypassing the formation of the tuber. This experiment is successful only with weak plants which cannot store up enough translocatable material to be able to produce a normal daughter tuber at their base. All these examples show that variations in the strength of the inhibition are responsible for changes in the normal polarity and rhythm of plants. On the one hand growth occurs not at the poles of the leafy stem but in the middle, and on the other hand during the course of development the rest period of the plant is eliminated.

It has already been mentioned that on a defoliated apical bud, growing at the tip of a rested tuber of Scrophularia, lateral buds will develop to replace the lost leaves. These buds will grow most actively on the node where normally the leaves would have grown most. It follows that the inhibitory factors which prevent the growth of lateral buds in this plant do not originate in the apical bud itself, but rather in the leaf primordia and in the material translocated from the mother tuber and the roots. These together cause the dominance of the apical bud. This is why removal of the leaves causes the apical dominance to disappear in sprouting tubers kept in the light. In identically treated tubers in the dark, the internodes of the apical bud elongate, but the lateral buds remain dormant even in the absence of leaves. The apical bud alone shows continuous growth, even though at first it is very small. In the dark, therefore, the leaf primordia lose their inhibitory effect; the small remaining apical bud becomes dominant over them and inhibits the growth of the lateral buds. The balance of correlations between leaves, buds, and even internodes is thus different in light and in dark. Considerations such as this help to explain the abnormal growth shown by so many plants in the dark, when the internodes elongate remarkably while the leaves do not even unfold. It is clear that this is basically a correlation phenomenon, for we can induce typically etiolated plants to bear normal-sized leaves, if we remove all the buds and all the leaves, with the exception of one primary leaf, from bean seedlings (*Phaseolus multiflorus*). The one primary leaf will then reach at least the size which it normally reaches in the light. If the whole experiment is carried out in the light it will be even larger than normal. In the light, then, each leaf is substantially inhibited by other leaves on the plant.[8]

Etiolated plants transferred to light at once stop their rapid elongation, and after a time their leaves enlarge, especially near the tip. These leaves then exercise a strong inhibition over the growth of the axis above them, which therefore slows down, while axillary buds grow out quite noticeably. Branching is therefore first determined by the exposure to a stronger light, which weakens the correlative dominance of the growing apical bud over lateral buds.

Coniferous trees constitute a classical example of apical dominance. According to L. Errera's explanation of it in spruce (Picea), it is not merely a question of food substances being used up by the apical bud at the expense of lateral branch primordia, but there is a special dominance factor in the tip which influences the lateral branches. It is commonly known that in spruce the apical bud gives rise to branches symmetrically all round (radially), while lateral branches

[8] Cf. the inhibition exerted on one another by the successive leaves of *Aster multiflorus* (Goodwin, 1937). The experiment described here, however, is quite new. (Ed.)

branch out on either side and maintain a more or less horizontal position. Only if the apex is damaged or broken off do lateral branches turn upright; in this case one of them usually soon prevails over the others, becomes vertical, and begins in its turn to branch radially. After several years such a replacement of the apex can hardly be detected from the side. If we ring the tip of the shoot, some of the branches will start to turn upward. The correlative influence is therefore propagated through the cortex, especially through the phloem.

As in the seedling bean plant (*Vicia faba* or *Faba vulgaris*), in which it was first demonstrated by Thimann and Skoog (1933) by replacing the main apex with indoleacetic acid, the correlative influence is a growth substance which spreads from the top downward. This transport results in the inhibition of lateral buds in the axils of leaves and cotyledons, and in spruce probably causes the horizontal growth of lateral branches and their bilateral branching. The ringing experiment can be carried out also with Araucaria, but in this case the lateral branches do not turn upward, they stay in the horizontal position but elongate much more actively. More often, however, both in coniferous and deciduous trees, the branches do not turn upwards after the loss of the main apex, but instead they produce on their upper side new vertical shoots.

The whole problem of these seemingly different types of apical dominance is epitomized in the experiments of Gunckel, Thimann, and Wetmore (1949) with *Ginkgo biloba*. The short lateral shoots of this tree are prevented from elongating by the terminal shoot, because they produce auxin for a very short time only. But their leaves enlarge and they produce some auxin; they develop into long shoots when the terminal bud is removed. This latter can be replaced by the exogenous auxin naphthalene acetic acid, which prevents elongation and thus maintains them as "short shoots."

The differences between the main axis and the lateral

branches are reflected in their anatomical structure, which is much simpler in the lateral branches, especially in conifers. While in the main stem, growing upright, the seasonal rings are concentric and equally developed on all sides, in lateral branches they are excentric, because a great deal more wood is formed from the cambium on the lower side of the branch; incidentally this wood, especially after soaking, is red-brown and pressure resistant. This red wood (rotholz) has thick-walled tracheids aligned spirally in single layers and with no inner unlignified layer. It is very different from the white wood on the other side of the annual ring which more nearly resembles normal wood. The lower, red wood is richer in lignin, it expands more easily, and its elements become twisted. There is an inner tension which tends toward equilibrium; this explains why, when such trunks are being cut down or sawed, they tend to warp or even to split. The same is true of the boards obtained from them.

Red wood, also called reaction wood, is caused by the accumulation of auxin in the lower side of the branches. This was demonstrated by applying on lateral branches indoleacetic acid, in the form of lanolin paste. In each case red wood was obtained directly below the paste in spite of the action of gravity. Gravity is actually an important factor in the location of red wood. Red wood forms in the main trunks if for some reason they become obliquely tilted; the lower side always has thicker annual rings when red wood forms. However, the tip of a straight trunk can also determine the position of red wood in lateral branches, sometimes even against the force of gravity, for it predominates on the abaxial side of the branches.

These phenomena in conifers and other woods can be compared with similar relationships between the main axis and the lateral branches of herbaceous plants. The so-called *lateral anisophylly* is here especially conspicuous — *adaxial leaves* (bending toward the axis) on the upper side of lateral branches being smaller than the *abaxial leaves* (bending away from the axis) on the lower side of these branches. If transverse leaves alternate with these anisophyllic pairs they are asymmetrical, one half of the blade being larger than the other. These asymmetries and variations of size are governed by inhibitory influences proceeding mainly from the growing apex, which inhibits the development of leaf surfaces on the upper side of the branches, while the opposite leaves on the lower side are, on the contrary, stimulated by the root system, which is closer to them.

It is actually possible to reverse this anisophylly, at least on the topmost branches, by removing the tip while root activity is still very intensive. The leaves on the upper side of these branches will then grow larger than those on the lower side. Such behavior is particularly clearly seen on isolated segments of the stem of Scrophularia, if they are planted in soil after they have produced roots and one supporting leaf has been removed. The two axillary buds left on the node then grow unequally, the one in the axil of the amputated leaf becoming larger than that on the opposite side. The more inhibited of the two shoots has smaller leaves on the side facing the supporting leaf than on the side facing the axis. The more rapidly growing shoot in the axil of the amputated leaf shows a normal anisophylly, with predominance of the abaxial leaf (Fig. 18b). This is obviously due to the inhibitory influence of the remaining supporting leaf, since on the two opposite shoots the leaves that are turned toward it are smaller than those turned the other way. It may be that these larger leaves on both axillary sprouts are stimulated by the activity of the roots which have been regenerated on the isolated segment.

A number of plant movements are related to this phenomenon. If we remove the apical bud from a rapidly growing leafy stem, then the topmost remaining leaves will turn upright, owing to more active growth of their lower sides. Such "hyponasty" is especially conspicuous if we cut off one of an opposite pair of leaves; the remaining one then bends over the apical stump to the opposite side. This reaction can be prevented by applying auxin paste to the cut surface of the stem, for auxin detaches itself from the translocating material and spreads to the lateral organs, in this case to the young leaves, which then grow less on the upper side because it is closer to the source of inhibitions. As a result the growing leaves execute a reverse movement called "epinasty," in which leaves which are at first vertical become horizontal, a position more appropriate for photosynthesis.

That these leaf movements again are caused in the plant by a gradual increase in inhibitions from the tip down is proved by young stems growing in the spring from the tubers of *Scrophularia nodosa*, which demonstrate these movements on their leaves and on the lateral shoots which grew in replacement of these leaves. Such lateral shoots do not grow upright as might be expected, but turn to a horizontal position, as their supporting leaves would normally have done. Only after further elongation do they become vertical and thus replace the defoliated stem which stopped growing. Inhibitions in a plant always increase greatly after the loss of the roots, so that if we remove all the roots from the mother tuber, the leaves bend downward with a gradual epinastic movement, as they would do on a klinostat when the action of gravity has been eliminated.

Similar phenomena occur in other buds such as those of potato; the variety *Fram* is particularly well suited. When the apical bud is kept intact, lateral branches (which develop from large buds) bend upward in a hooklike structure, because the tip inhibits the upper side of these branches. If the tip is removed, then the ends of the branches start to bend down, because the nutrient material coming from the tubers and roots stimulates the growth of the upper side of the sprout, which is thus no longer inhibited (Fig. 12a). This hooklike movement can be reversed permanently in pea seedlings by applying auxin paste on the convex side of the plumule (Fig. 12b, c).

Even the direction of branches, both in herbaceous and in woody plants, is determined by factors originating in the apical bud and in the roots, respectively. On lilac bushes which have been cut down a great deal, the new shoots are only slightly divergent, almost parallel, while on normally kept bushes the shoots form wide angles between them. Weak shoots form much wider angles than stronger ones. We have found that if the scales were removed from the flower primordia at the beginning of their formation, the shoots then grew completely parallel; this shows that relatively weak inhibitions exist in the tip of the flower stalks at the beginning of summer.

In all these cases several growth hormones act together, and the application of auxin can change the angle of the branches in the same way as an appropriate cut. This has practical importance for the control of the disposition of branches, especially in young fruit trees. Many monopodially branching deciduous trees will straighten up their lateral branches if the tip is removed, and if then the tip is replaced by auxin paste the branches will regain their original position. Similar conditions are encountered in leaves[9] and in runners which grow at the base of leafy stems and serve for the propagation of the plants, either directly or by forming a tuber at their extremities.

The lower branches of *Circaea intermedia* are actually runner-like shoots; they can become vertical at the tips if the stem above them is cut off or covered with dark paper. Similarly, potato stolons, even if they have already started to form tubers, will bend upright after the loss of the main stem and grow up from the ground in the form of leafy stalks (Knight,

[9] Especially useful for these experiments are herbaceous plants, such as certain varieties of *Impatiens royeli*.

1806). Historically, this was the first correlation in plants to be proved.[10] A similar change in geotropic reaction occurs after prolonged rainy weather, so that here again a decrease in inhibition can be related to the influence of the root system, which is no longer inhibited by the stem and can thus actively draw in and assimilate mineral substances, causing rejuvenation of stolons and tubers. A similar effect can be obtained by spraying the plant with gibberellin, which, if used at the right time before harvest, can substantially shorten the rest period of the tubers. On the other hand, we cannot expect to promote the development of tubers by applying the usual inhibitors, such as maleic hydrazide, because this only prolongs their rest period.

The practical significance of indoleacetic acid and other substances with similar physiological effects in relation to these correlations cannot be denied. They can be used with great profit, for example, when a rainy autumn follows a dry summer and there is danger of early germination of buds on potato tubers, or of premature fruit drop on apple and pear trees. This may be owing to the sudden outburst of activity of the root system, which stimulates the growth of lateral vegetative shoots and thus indirectly inhibits any further development of fruits. The use of the above-mentioned substances is especially profitable in the culture of cotton, in which producduction can be increased up to 30 percent.

Abscission processes, frequent in the last phase of plant development, can be greatly delayed by treatment with these substances. Such responses also show the significance of integration in plants, since even an apparently simple process like the falling of leaves may depend on other organs in the plant, on the growth of the tip, the production of flowers and fruits, and the function of roots. Variations will be observed accord-

[10] The many observations of Duhamel du Monceau, showing that buds cause swelling of the cortex below them, especially just above a ring, are much earlier (Physique des Arbres, Paris, 1758). (Ed.)

ing to whether the roots are growing or not. The correlation between a leaf and the abscission layer at the base of its petiole is similar to that between a leaf and its axillary bud. This means that if the leaf is still fresh enough to produce an abundance of stimulators or of their precursors, the axillary bud does not fully develop, and the abscission process (which begins as the disintegration of pectic lamellae between cells of a special tissue at the site where the petiole joins the stem) does not take place.

In practice auxins are commonly used to prevent the premature drop of healthy but unripe fruit, which can increase the harvest a great deal or, if needed, delay it to a more appropriate time. A good and relatively cheap product is 2,4,5-trichlorophenoxy acetic acid or one of its salts. In some cases where, on the contrary, there is a risk of periodic fruiting, especially in certain varieties of apple trees, these and other products can be sprayed on the unfolding inflorescences, to decrease the number of prospective fruits and check their detrimental influence on the formation of flower buds for the following year.

The formation of tubers, bulbs, and rhizomes is also important for plant integration. Such organs ensure vegetative reproduction and thus the preservation of the life of the plant during unfavorable, cold, or dry periods, while in perennial plants they have other advantages. The formation of most of these structures requires specific conditions. Tubers form best in the dark and in a somewhat humid environment, so that the optimum condition for their development is with the base of the stem covered with soil. If the underground bud primordia are removed, tubers can develop even above the ground, though not so well. In the potato plant tubers have even been obtained at the top of the stem by keeping the plant continuously in the dark (Fig. 12b). The axillary tubers formed in the light are much smaller, and are usually terminated by a large bud composed of dormant primordia of folded leaves. Even when underground tubers are already formed, their diameter can be reduced by removing part or all of the stem. As soon as the stem regenerates, the ends of the tubers thicken again, so that by repeating this experiment one can obtain a tuber with a succession of knots, like the beads of a rosary — proof of the correlative interdependence of tubers and aerial parts (Molotkovsky, 1954).

In many plants the formation of tubers depends to a great extent on the length of day, which is also evidence for the integration of the plant. Among our experimental plants, *Circaea intermedia*, under short-day conditions (8 to 9 hours' day-length), forms only short rosettes without any trace of flowers, and soon afterwards small tubers develop underground on short runners. Certain varieties of potatoes growing wild in tropical South America and brought to our country for experimental purposes show the same phenomenon. Thus *Solanum leptostigma*, *S. semidemissum*, *S. antipovieczi*, and others form tubers under 10 to 12 hours' day-length but do not form any during the long days of moderate or cold latitudes.

Tubers of *Circaea intermedia* are noted for their deep dormancy, during which, under favorable conditions of temperature and humidity, they can elongate at the tip but form only short, thick internodes with tiny scales. In this way they continue their tuberal growth, and as a result the reserve contents at their basal end gradually get used up. Tubers which have gone through part of their dormancy may form a very thin vertical stem, which soon bends at the tip, assumes a horizontal position, and finally becomes a runner, covered in the light with scales transformed into small leaves. The tubers become exhausted by the formation of the runner, and the normal leafy stem usually does not grow out.

It is interesting to note that only a typical leafy stem, growing from a completely rested tuber, brings about the formation of cork in the maternal tuber, through cell division in the pericycle. Such an alteration of tissues depends on the growth of leaves, at least until these reach 7 to 8 mm in length. For this reason it never occurs on plants of which the stem has been transformed into a runner.

If young plants are placed in an ethylene-containing atmosphere, they will produce terminal and axillary runners, even if they have gone through the whole rest period. The runners bend downward and can even thicken into tubers. This shows an increase of inhibitions, which is always necessary for the formation of runners and even more so for that of tubers on these runners. Ethylene, which is the most important component of illuminating gas (as far as its influence on plants is concerned), causes sensitive objects, such as pea seedlings or young potato sprouts, to thicken at the expense of their elongation growth. If, however, tubers are to be formed then these inhibitions do not act merely quantitatively, but they bring about a deep transformation of substances in the plant.

The same concepts can be applied to the well-known experiments of Razumov (1931), professor at Leningrad. He used tuber-forming wild plants such as *Ullucus tuberosus* and *Oxalis tuberosa*. When the leaves were kept in short days, their axillary buds elongated into positively geotropic, thin shoots which penetrated into the surface of the ground and thickened into tubers (Fig. 22). If, however, the leaves were then exposed to long days, the tubers elongated, became negatively geotropic, and transformed at their extremities into leafy stems.

Some experimenters, such as Hamner and Long (1939), have tended to think of a hormonal substance necessary for the development of tubers, which could be formed in the leaves only in short days. Such a hypothetical hormone would correspond to one of the organogenic substances of Sachs (1880). It is, however, difficult to believe that one single chemical compound could cause such drastic changes in the development of the plant. It could not be so even for such simple cases as tuberization of the dahlia, which is also stimu-

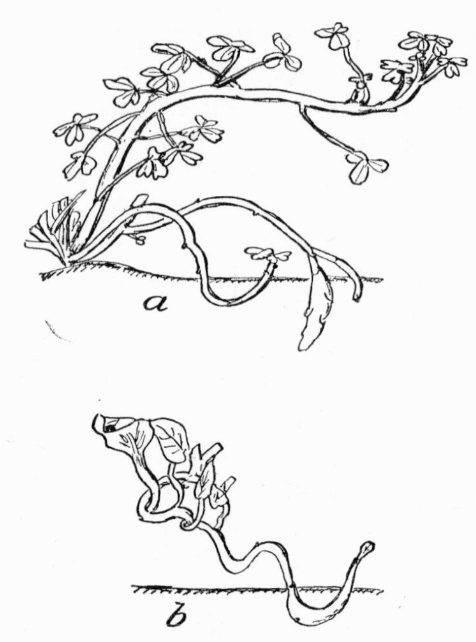

Fig. 22. *Oxalis tuberosa:* (*a*) at first, in short days, runners grew into the soil and partially swelled into tubers, but later, in long days, the runners curved upward and formed leaves. Ulluco (*Ullucus tuberosus*): (*b*) tubers, formed in the soil during short days, elongated later above ground in long days (after Razumov).

lated by short-day length. Tuberization must be thought of rather as an adaptation of these plants to the alternation of long- and short-day lengths in nature, which involves at the same time an alternation of cold and warm weather and is therefore closely related to the plant's periods of rest and activity. This concept, however, does not supply a causal explanation of the formation of tubers, especially as compared to the development of the leafy stem in the case of the axial tuber, or to that of the absorbing root in the case of those tubers originating from roots.

In the case of the axial tuber, correlations between the axis and the apical bud are altered, so that even a slowly growing apex inhibits the growth of axial internodes and the elongation of lateral buds. During the normal growth of a leafy stem the apical bud, which is the site of activation of auxins and probably also of gibberellins, promotes the elongation of axial internodes, but during the formation of tubers it acts as an inhibitor, so that internodes do not elongate but thicken greatly. Evidence for this interpretation was derived from experiments with isolated sections of the stem of *Scrophularia nodosa*, taken from the basal part where the formation of tubers predominates. A longitudinal cut through the axis separated the two leaves of each pair, and the terminal meristem of one of the young axillary buds of each node was removed. The intact control bud enlarged only very little, through an almost imperceptible thickening, while the decapitated bud became transformed into a round tuber (Fig. 23c). There is

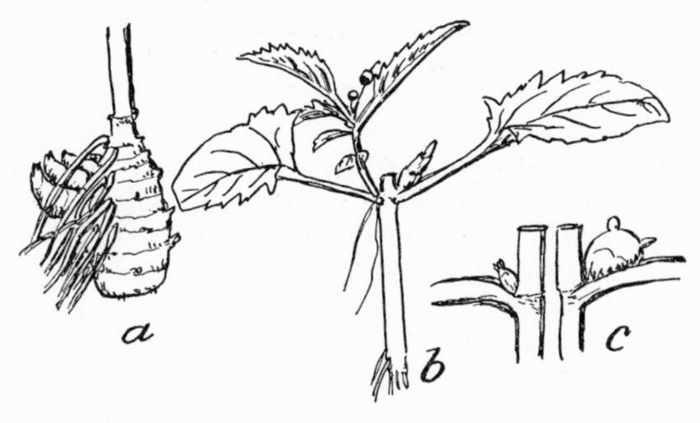

Fig. 23. *Scrophularia nodosa:* (*a*) mother tuber, from which roots have been continuously removed on one side, forms daughter tubers only on the side where roots remain; (*b*) defoliated axillary bud becomes a tuber, while the opposite foliate bud forms stem with leaves and flowers; (*c*) decapitated axillary bud becomes a spherical tuber, while the opposite intact bud remains inhibited.

no doubt that, during growth, inhibitions increase in the plant. The growth of leaves is thus slowed down and assimilation products accumulate. The storage of these products in appropriate places is favored by the limited growth of buds.

That inhibitory factors become concentrated in the apical bud can be demonstrated with tubers of *Circaea intermedia*. Tubers of which the apical bud has been left intact remain dormant for a long time without any change, while tubers from which the apical bud has been removed immediately start developing lateral runners, from the axils of scales which were left at the apical end. It is indeed remarkable that the intact tuber does not grow, while the decapitated one brings forth runners from lateral primordia; these had previously been inhibited by the dormant bud (Fig. 24a and b).

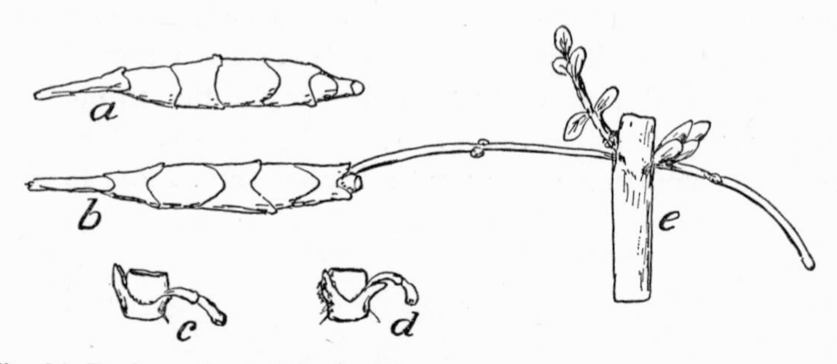

Fig. 24. Enchanter's nightshade (*Circaea intermedia*): (*a*) resting tuber does not elongate and lateral buds are inhibited; (*b*) resting tuber with tip removed forms long runner; (*c*) on tuber segment, bud in axil of the excised scale develops; (*d*) if both scales remain and one is treated with auxin paste, the bud grows in the axil of the other. Lilac (*Syringa vulgaris*): (*e*) on resting segment from the middle part of a branch, bearing winter buds of unequal size, kept at room temperature, the smaller bud begins growth while the larger one remains dormant.

Again there is a certain analogy with the shoots of woody plants, for example, those of lilac, which do not sprout when transferred during their rest period into a warm room, as long as the top large buds remain intact. If these are removed, or the shoot is sectioned below them, small, previously dormant buds start to develop. These small buds are thus not in a true rest period, but are merely inhibited by the large upper buds. It is only the latter which can enter into a deep true dormancy.

The various depths of dormancy of large and small buds

can be shown on isolated single-node segments of a lilac branch, situated between nodes bearing large growing buds and those bearing dormant buds, and on which the two opposite buds develop very unevenly. In isolated cultures of these segments a small, previously inhibited bud may grow more intensely than a large bud, which often remains closed for a long time (Fig. 24e). The difference between this situation and that seen in the tubers is only the fact that in woody plants the reserve substances are stored away from the buds, while in tubers they lie between the buds. From the point of view of plant integration one must not overlook the resting states of the plant, for theoretically they are as interesting as the more dramatic phases of vegetation and reproduction.

Two examples will show especially clearly how the formation of tubers depends on products formed by the leaves: one is the spherical root of the sugar beet (*Beta vulgaris ssp. saccharifera*), the other, the aerial tuber of the kohlrabi (*Brassica oleracea var. gongyloides*). Vöchting (1908) studied in detail these two cases and found that each leaf is responsible for the growth of a particular section of the tuber. For if, while the tubers are developing, we remove the leaves of the plant on one side, then the tuber stops developing on that side. Since one side thus develops much more than the other, the top of the tuber appears to bend down on one side. In the case of the sugar beet, the growth of additional rings of vascular bundles is much more active on the leafy side of the plant. In this respect the effects of young growing leaves differ from these mature, actively assimilating ones. Podešva (1947) removed from one side of the tuber only mature leaves, retaining the youngest ones, while on the other side he retained the old leaves and removed the young ones. The tuber grew more on the side bearing the young leaves. The explanation probably is that young leaves supply a higher concentration of growth factors (especially auxins) than the mature leaves, while the latter would tend to inhibit growth by their assimilation products. It is amazing that in spite of the complicated network of vascular bundles, as Vöchting (1908) himself pointed out, the differences between the two sides remain perfectly defined, so that evidently such influences do not spread transversely across the tuber. The same is true for the sugar beet, whose young leaves stimulate the growth of the tuber in thickness, while the mature leaves raise the level of sugar in it. For this reason, the growth of the root is promoted by the removal of mature leaves, but at the expense of the sugar content. One has therefore to be very careful about which leaves to remove and which to retain.

Similar conditions exist undoubtedly even in reserve organs, which are not, like these two, on the main stream of substances from the root to the leaves and back to the roots. For example, tubers of *Scrophularia nodosa* also need for their growth a certain level of stimulation, as is shown by plants grown from maternal tubers, planted in the spring vertically in the soil, so that gravity will not determine the localization of roots on these tubers. Normally tubers are formed laterally at the base of the stem and at the tip of the tuber of the previous year, on the lower side where many more roots are formed than on the upper side. On vertically planted tubers the roots can very easily be removed on one side so as to cancel the influence of gravity. On the other side, then, the roots cause an intense production of tubers, which therefore need a certain degree of stimulation supplied by these unilateral roots (Fig. 23a).

One can easily infer that the presence of the roots at the base of the stem is an important condition for the formation of tubers at that point, because it is only there that roots decrease their inhibiting action in favor of the growth of new tubers. There is one particular node on which the tubers are always largest, and their size decreases with the distance away from that node, just as the size of the leaves decreases from the center of the stem to the poles. At the time when daughter tubers are formed they represent the center of inhibition, so that all other primordia on the plant become arrested in their growth. This affects not only the axillary buds of the leafy part of the stem but the very top parts also, with the inflorescences. An interaction between the formation of flowers and of tubers can be shown by illuminating the base of the stem; this has a negative effect on tuber formation. Darkening the inflorescence, on the other hand, retards the formation of tubers, while the inflorescence grows a meter high and new flowers are continuously produced at the expense of the leaves which remain in the light.

Lilies, such as *Lilium candidum* and *Lachenalia luteola*, do not form seeds under normal conditions, or only very rarely. They can be forced to do so, however, if the floral axis is isolated from the bulb and set in water.

Often the relationships between the organs of vegetative reproduction, on the one hand, and flowers and fruits on the other, are not so simple. In the case of *Ficaria verna* and *Dentaria bulbifera*, for example, one cannot increase the production of seeds by removing lateral tubers or bulbs. Ficaria is sometimes referred to as sterile, its sterility being largely compensated by the production of great numbers of tubers, not only at the base of the rootstock but also in the axils of leaves. However, our experiments show, on the contrary, that the greatest number of fruits is obtained in locations where the tuber production is also stimulated. Fruiting depends, therefore, on active development of leaves and branches, and this in turn (especially in good garden soil in which Ficaria is especially fertile) is determined by development of the roots. On the other hand, in poor sandy soils, or in acidic soils, Ficaria forms only a small root system, weaker and shorter stems, rather yellowish leaves, and almost always sterile flowers. There is also a difference in the shape of the tubers: fertile plants have elongated, sticklike tubers; sterile plants have usually round or egg-shaped ones. The shapes do not indicate two different varieties, for if we transplant these plants from one habitat to the other, the one type is rapidly transformed into the other. The degree of fertility is more likely determined by stimulations and inhibitions which depend on the degree of development of the root system. Vigorous development of the root system stimulates the plant, as was shown before in flax seedlings. A deficiency in the root system can cause the drying up of flower buds, so that only the formation of tubers takes place, and even this is also limited. Thus the plant is kept alive and preserved for subsequent generations, even though it is very weak and cannot develop flowers or fruits.

The reverse phenomenon occurs in annual plants which flower with great regularity, and show a direct relationship between the production of flowers and fruits and longevity. If we remove early enough all flower primordia from annual, short-lived plants, such as *Reseda odorata*, the plants will remain alive for years and eventually form shrubs. This does not apply to all annual plants, since more commonly, when flower buds are removed, inhibitions arise leading to the destruction of the plant.

Even in perennial plants, vegetative growth can be promoted by removing flowers. In the hyacinth, the flower stalks are regularly removed in order to increase bulb development. Spraying with MH is also recommended, since it inhibits flower formation, and the products of assimilation are then used for the growth of bulbs. In this method there is no danger of fungus infection, which may occur when the stalk is cut off. There is also an increase in the production of tubers by *Scrophularia nodosa* if the flower primordia are removed. Thereafter the assimilation material produced, even that from the uppermost leaves, flows into the base of the stem and is used for the growth of tubers. If, however, the upper part of the plant is isolated, then the same nutrient substances are used for the formation of flowers.

All of this is again important in connection with the integration of the plant. When the plant is intact its growth conditions are different from those established when it is divided into parts. The aerial system is interconnected by very complex interactions between individual organs. Roots and leaves have mainly a supplying and regulatory role, while growing organs, such as buds, tubers, and flowers, which consume the products of roots and leaves, have an opposite action on the main organs of plant nutrition

The development of roots involves simpler correlations. The growth of roots depends to a large extent on organic material obtained from the aerial parts, principally the leaves. However, external factors, such as the temperature and the composition of the soil, influence the roots directly, and the root system can be considered as the most plastic part of the plant. In pea seedlings, for example, the growth and branching of the roots depends on the cotyledons. If one of the cotyledons is removed from very young seedlings, lateral roots develop more quickly and in greater number on the side of the intact cotyledon. This must be due to the direct influence of the nutrients and of specific substances, such as growth factors and vitamins, produced by the neighboring cotyledon. The branching of roots can be increased by soaking the seeds, prior to germination, in a solution of auxin. If we remove one of the cotyledons on the plants obtained from these seeds, so many roots are formed below the remaining cotyledon that they fuse into disclike fasciae. Later this specific action of the cotyledons is supported by stimulators and vitamins, supplied to the roots by the aerial stem.

The requirements of the roots are well known, thanks to the culture of isolated roots in artificial media, which contain, in addition to mineral salts and sugars, special substances such as thiamine, lactoflavin, and biotin. The production of these substances in the cotyledons is not unlimited, so that if the main stem and all the successive cotyledonic buds are systematically removed from a young plant, the branch roots develop in short zones separated by large bare patches. Up to five such branching zones can be found on the main root, while the tip end of the root remains bare over a relatively long stretch. The inhibitory influence of the tip of the root prevents lateral roots from growing out until a certain distance from the tip is reached.

Inhibitors have been identified in the root by Libbert (1954), and similar inhibitory substances contribute to the pattern of growth of lateral branches. It can easily be imagined that the cotyledons, without the aid of other aerial parts, especially of the leaves, can supply only limited amounts of the substances necessary for branching. The exhaustion of these substances would interrupt the branching process.

Inhibitions in the roots can be increased by treatment with external inhibitors. The plants which develop from seeds previously soaked in maleic hydrazide form a root and a stem as long as they can activate necessary substances for their own growth. Then when the root becomes 5 to 7 cm long its growth stops, its top part enlarges into a sort of tuber, and there is no branching of the resulting spherical root, although the cotyledons are not yet exhausted. On the other hand, the petiole and the cotyledonary node thicken a great deal. The plumule also grows very little, and the internodes between scales and leaves do not elongate. In a few instances a "replacement root" may appear on the hypocotyl near the cotyledons and will form branches. Such branches appear only at a certain distance from the point of insertion of this new root, which again shows a strong inhibition in the whole plant. The growth of the stem is not necessarily stimulated by the appearance of this new root, and vice versa, the stem in some cases may resume its growth after a temporary arrest, without the root's showing any change.

This example shows some independence between the development of the root and that of the stem. Such independence is seen in those embryonic plants which have a rich reserve content in the seed, as, for example, in corn. It can be explained by the ability of both these organs, at least during the embryonic stage, to draw nutrients from the cotyledons (or from the endosperm) more or less independently from one another. The independence may be increased by applying substances which selectively stimulate the growth of certain organs. Seeds given an overdose of auxin develop large numbers of roots, while the growth of the stem may be completely inhibited. On the other hand, triiodobenzoic acid stimulates the growth of the stem more than that of the roots. A similar effect is obtained with gibberellin, which results in an abnormal elongation of the stem at the expense of the root. Triiodobenzoic acid often causes an overproduction of flowers, by stimulating branching. In some plants, however, concentrated solutions of triiodobenzoic acid have inhibiting action and may cause malformations. Members of the Solanaceae like *Solanum lycopersicum* (tomato) and *Capsicum annuum* (capsicum) or of the Compositae, like *Helianthus annuus* (sunflower), react to triiodobenzoic acid by increasing their flower and fruit production, while flax and plants of the Bryophyllum group form only funnel-shaped structures which result from the fusion of leaves of a certain age. The growth of the stem is thus arrested for a while.

Insufficient light also limits the growth of roots, short days having this effect while long days stimulate root development. That production of flowers is favored in some plants by short days is partially due to the adaptation of short-day plants, which are usually of tropical origin, to their habitat, and partially also to the insufficient development of their root systems. That a weak development of roots contributes to a greater production of flowers and fruits is a well known fact, commonly applied in fruit growing.

Just as the development of roots depends on the aerial parts, so the roots in turn influence the development of the stem. A fine example is given by Chailakhian's (1958) experiments with the long-day plant *Rudbeckia bicolor*, which does not flower in short days and will not flower in long days either, if its roots are removed. But it is sufficient to expose one of the leaves or even one eighth of a leaf to long days in order to make the plant flower. The plant can also flower in long days, even if it has no roots, provided a small amount of borax or boric acid is added to the water. The microelement boron can therefore replace specific root substances which stimulate the growth of the flowers. However, borax will not make the short-day rosettes flower.

A large number of factors, among them especially nitrogen in appropriate chemical forms, can influence the relationship between the development of the root and that of the aerial system. If the leaves are highly stimulated by an excess of nitrogen this inhibits the formation of flowers. This fact led Klebs to his well-known view of the importance of the ratio of carbohydrates assimilated in the leaves to the nitrogenous substances taken up by the roots. A high value of this ratio leads to flowering, a low one to vegetative growth. It is not valid, however, for plants whose flowering is controlled by day length.

In order to explain the observed correlations, particularly those between the roots and the stem with leaves and flowers, it is necessary to study the migration of many substances. This can be done conveniently with the aid of radioisotopes. As was shown in very simple experiments with flax seedlings, the movements of specific substances which control the growth of organs can be governed by the movement of common foods, such as carbohydrates, amino acids, and mineral salts. Their movement in the plant is facilitated by special conductive tissues. Organic substances, especially, move through the phloem, composed mainly of sieve cells, with a definite speed (in the case of sugar, 1 meter per hour). In the phloem as a whole the flow of organic substances can take place in opposite directions at the same time.[11] The mechanism of these movements is not yet satisfactorily elucidated, but some authors consider that the cytoplasm itself participates in the streaming, depending on the physicochemical conditions of the cells.[12] Maximov (1946) and Kursanov (1952) were among those who discovered these movements.

Sabinin (1957) found during the course of his studies of the migration of indispensable mineral elements, N, P, and S, that there are two circulations in a plant, a long circuit which goes from the leaves through the axis to the roots and back through the axis to the leaves, and a short circuit which is established between each leaf and the main axis. Experiments show that correlative influences can spread from one organ to another even if there are no conductive tissues, because cells are in communication with one another through cytoplasmic bridges, the plasmodesmata, which connect all the living parts of the cells of the whole plant into one continuous protoplasm.

The organization of the plant thus shows some resemblance to the nervous system of animals, although there are no real nerves. What is sometimes called the nervature in plants consists only of conducting tissue. The most conspicuous elements of this are bundles of vessels, basic components of the xylem, and also mechanical elements, both of which types constitute a skeleton for the plant; without these elements the more or less liquid living matter could not withstand the unfavorable external factors to which it may be exposed for hundreds of years. There is no structure in plants that could be compared to the brains of animals, and nothing quite analogous either to animal hormones, since auxins act more as stimulators in a general sense than as hormones with specific effects. At the

[11] Probably not the case in individual sieve tubes. (Ed.)

[12] Mature sieve tubes contain very little cytoplasm. Most Western authors now consider that the solution in the sieve tubes moves en masse. (Ed.)

same time their regions of origin can in no way be compared to the specifically localized endocrine glands of animals.

Among the other persistent tissues, the phloem, comprising sieve tube elements and their companion cells, seems particularly active. The role of phloem is not only to conduct organic substances, including large molecules which can spread through the plant via the rather large openings in the vertical walls of the sieve cells, but some also consider it as replacing the endocrine glands of animals, though such glands would be scattered throughout the plant. Němec (1932) observed in the tumors formed on the sugar beet after infection by *Heterodera schachtii* that individual sieve cells were scattered around the enormous cell fusions. These cells could not have functioned as a conducting system, but seemed to be contributing to the formation of the tumor. Haberlandt (1913) hypothesized that the phloem (leptom) produces special hormones (leptohormones). A little cube cut out of the potato tuber in such a way that it contains no sieve cells will not form a layer of scar tissue (wound cork) on its surface, unless a little particle of the tuber containing some phloem is added to it. In our experiments a section of lilac would flower remarkably rapidly, when its base was set in water and all the wood was cut out while the bark tissues were retained. The phloem then became very lush (Dostál 1941).

Each living cell is capable of activating auxin and transforming it into other forms, which may be active, inactive, or inhibitory. Intensely growing parts, namely dividing tissues or meristems, are in this sense most active. Since auxin is indoleacetic acid, it could be thought of as a by-product of protein metabolism, being derived from tryptophane. According to recent experiments, substances highly active in promoting elongation, namely gibberellins, are formed in higher plants, and are not (as at first thought) limited only to lower plants. For a long time both gibberellins and auxins were believed to be products of fungi and could be obtained from the substrate of the mold *Rhizopus suinus* in the case of auxin and from that of *Gibberella fujikuroi* in the case of gibberellins. These affinities between such distantly related organisms show the great importance of specific substances throughout nature. Indeed, many lower plants, even without cellular structure, may possess a complex plant body which simulates the organization of the higher plant. For instance, in the alga *Caulerpa prolifera*, the creeping rootstock, often 1 m long, grows upward into vertical, leafy structures, called assimilators, more than 10 cm tall, and downward into filamentous, finely branched rhizoids (Fig. 7a). The green assimilators correspond to the leaves of higher plants and the rhizoids to the roots.

The entire alga is filled with a continuous protoplasm, containing numerous nuclei and deep green chloroplasts embedded in a clear cytoplasm. The protoplasm covers the structural framework which supports the alga by forming a lining to the thick outer membrane which protects the living matter against outside influences. The protoplasm circulates for the most part through the whole plant in a number of streams of variable strength.

In spite of the lack of cell structure the alga develops very regularly: the rootstock elongates at one end and produces, at regular intervals, oblong assimilators which subsequently enlarge their surfaces. On the lower side, and at shorter intervals, the rootstock forms finely branched rhizoids. The rear end, bearing the old assimilators and rhizoids, dies off continuously and the living content is transferred forward. There are no partitions except in case of injury, when the living part becomes separated from the dead part by the production of a coagulate of the vacuole content, on which a new cell wall is deposited.

The rhizoids of the acellular alga have the same role as roots in higher plants. They stimulate the growth both of the assimilators and of the rootstock. If a segment of the upper part of the plant is isolated, it is rhizoids which develop first;

and the intensity of their growth determines whether new assimilators or new rootstocks will be formed from the segment. That they should exert a stimulatory influence is to be expected, since they develop actively in the rich nutritive substrate of dead organic substances. However, it is more surprising that assimilators floating freely in water, or planted in sand, exert inhibitory activities. The inhibitory influence of assimilators originates in their edges, so that the new organs are as a rule formed as far away as possible from the edge. Only in plants grown in the dark for a long period of time does the upper border of the assimilator lose its inhibitory effect; then fine, needlelike shoots grow from the upper edge (Fig. 6a).

One of the most interesting phenomena is that if the alga is transferred from quiet sea water to an aquarium with running water all the living content from the assimilators migrates into the rootstock, so that the assimilators become completely bleached and the rootstock dark green, almost black. If such plants are returned to calm water numerous needle-shaped shoots arise from the rootstock completely irregularly, for the plant has no leaves with which to regulate the development of new organs. The entire content of this weird alga, when all its components, cytoplasm, nuclei, and chloroplasts, are mixed together and stuffed into the rootstock in an apparently structureless mass, is in some ways comparable to living matter in its primordial state, the simple oblong shoots resembling those of the oldest and most primitive plant forms. *Caulerpa prolifera* reaches this strange state at the end of each vegetative period, when reserve substances, especially starch, migrate into the rootstock.

There are no real differences between the requirements and the correlative influences of the organ analogues (organoids) of the alga and those of the true organs of higher plants. The integration of the plant is thus determined by the same correlations in both cellular and noncellular plants.

In *Caulerpa prolifera* we lack only the flower analogues. There has been a long search for the reproductive organs of this alga. Ultimately their existence came to be generally doubted, by analogy with some higher plants which have lost the ability to form seeds, while still actively reproducing vegetatively. (For instance, the flag, *Acorus calamus*, originally from southern climates, has persisted for hundreds of years in our waters, by vegetative reproduction of the rootstock.) In the course of our experiments, however, we have observed that in one night the whole living content of the alga can divide into countless gametes, each about a hundredth of a millimeter in diameter, containing one nucleus, one chloroplast, two flagella, and one eye spot. These gush out, before dawn, through special papillae formed the previous day on the assimilators, and through other openings all over the alga caused by enzymatic digestion of the membrane. Next day the rest of the plant decomposes and completely disappears. The sexual reproduction of this alga, which for so long escaped the attention of botanists because of its speed and because it takes place at night (although it is quite common at the end of summer), differs radically from the sexual reproduction of higher plants. In higher plants the reproduction affects only a small part of the plant body, while the rest may remain alive for hundreds and thousands of years, as in such giant trees as the Australian *Eucalyptus amygdalina* and the Californian *Sequoia gigantea*.

External factors act much more directly on the alga than on higher plants, with their complex differentiations of cells into tissues whose living matter is divided into smaller parts united only through the plasmodesmata. This structure contributes to their solidity, longevity, and especially to a rigorous division of labor. It is only in their embryonic parts, without thick membranes or dead cellular elements, that higher plants show some resemblance to the continuous protoplasm of the acellular alga.

In the alga *Caulerpa prolifera* an embryonic protoplasm forms from an assimilator or from a piece of the rootstock, in the shape of a vegetative tip, simulating the vegetative apices of higher plants. At first, as can be shown experimentally, these are quite indifferent, and only under the influence of other organs and of external factors do they give rise to the analogues of organs of higher plants.

In all multicellular plants also, embryonic development is modified by external and internal factors. Probably because of their complexity, embryonic development is determined to a large extent very early, before any sign of their organization appears. A good example is the climbing ivy, *Hedera helix*, characterized by two types of leaves and two different stem orientations. The young form has deeply lobed leaves on trailing stems with adhesive roots on the lower side, while the mature form has oval whole leaves on vertical flowering stems. According to Němec (1934) it is possible to change the shape of the leaves and the direction of the stem, and thus to obtain plants with trailing stems bearing oval leaves or plants with vertical stems and lobed leaves. The shape of the leaves is, however, always determined a half year, or even a whole year, before it appears on the elongating stem. Even though light intensity and humidity may influence the plant, their effect is not felt directly by the unfolding buds, but the differentiation of stems and leaves is determined by leaf products which influence the very young bud primordia. It is quite possible that special organogenic substances may intervene. At least one result seems to suggest this: in a combined culture of the mature form — with roots and entire leaves — and of the juvenile form, both in the same nutritive solution, the adult form began to develop in the same way as the juvenile form, and lobed leaves followed the simple oval ones. It is thus possible to rejuvenate the mature form.[13] The

[13] Gibberellic acid causes an even more complete reversal to the juvenile form (W. J. Robbins, *Am. J. Botany* 44:743–746, 1957; 47:485–491, 1960). (Ed.)

reverse effect is not readily obtainable, which may explain why the grafting of juvenile forms on mature ones, especially in the case of fruit trees, does not speed up flowering.

A close relationship exists between the young and the mature forms, which was observed by Michurin (1948) during his long years of assiduous work. He developed some observations about the chronology of correlations, which helped him in the selection of hybrids used to propagate fruit trees and other horticultured plants. His experience was especially valuable when choosing the hybrids in various stages of life. From the beginning of the germination of the seed, Michurin would evaluate the cultural value of the future variety by the size of the cotyledons, the strength of the hypocotyl, and sometimes the number of cotyledons. The color of the cotyledons already suggested the color of fruits, and in the case of roses even the color of the flowers. The second selection was made at the end of the vegetative period. At that time he selected strong individuals with large leaf blades, short thick petioles, thick ends of shoots with rounded, only slightly indented leaves covered with thick hairs. After the fall of leaves, he selected plants with large round buds, arranged close together on the axis, almost in circles, for such plants promised a firm flesh of the future fruit. On the other hand, a scattered, spiral arrangement of buds was considered to be a sign of loose flesh in the fruit. In the case of drupe-bearing trees, he used as his criterion rounded buds, arranged in groups of three, and deep grooves on the petioles. Dark color of the bark was considered a sign of late ripening of fruit, while light-colored bark foretold summer ripening. The third selection was made in the third year, when in addition to these criteria he could consider the quality of the fruits (see Chailakhian, 1955).

Němec, in his book *History of Fruit Growing*, discusses the distinguished Czechoslovakian grower J. E. Proche (born in Nový Bydžov 1822, died 1908), who improved many varieties of fruit trees and shrubs and gained time by evaluating (even before Michurin) the quality of the future tree by certain signs present in the youngest individual hybrid seedlings. His fifty years of indefatigable work unfortunately did not receive recognition by his contemporaries, and not even afterwards have his merits been as widely publicized as those of Michurin, the great Soviet recreator of nature.

Krenke (1933) used the same criteria to predict the fertility, precocity, and resistance of various cultured herbaceous and woody plants. In all these cases the relationships are certainly much more complex than in the growth correlations demonstrable experimentally. Their clarification, especially from the causal viewpoint, will undoubtedly contribute to the possibility of the transformation of plants. Even plant physiologists recognize that the greatest obstacle to such an undertaking is our incomplete knowledge of correlations and of their dependence on external factors, which affect them just as they have affected plants all through their phylogeny.

CHAPTER IV

Regenerative Mechanisms for Preservation of the Plant's Integrity

Plants, like animals, have developed to varying degrees the ability to replace missing parts, an ability which we call regeneration. This is an important mechanism for the preservation of plants, for since they are sedentary they are much more likely to be damaged than animals which can move about. Nevertheless some plants do not possess this ability, or at least possess it only incompletely, so that preservation of their lives is not ensured.

It has long been known that plants can be propagated by cuttings or by grafting and that the whole plant can be thus regenerated, either through the formation of new parts or by cooperation with organs derived from a partner. Let us first consider the leaves, which are often thought of as the most typical organs of a plant (Timiriazev, 1951, in fact considers the leaf as the real plant). Leaves are not always able to give rise to a whole plant after they have been isolated, even under the best conditions. This was demonstrated by the experimental work of Hagemann (1931) dealing with the regeneration of isolated leaves of a large number of species of plants. Out of 1196 species, only 308 formed both buds and roots. In 888 species the leaves sooner or later died, either because they did not regenerate any organs at all (389 cases) or because they formed only roots (499 cases), which only temporarily lengthened the life of the cuttings.

These numbers, however, do show that many leaves, the main organs of carbohydrate production, tend to complement

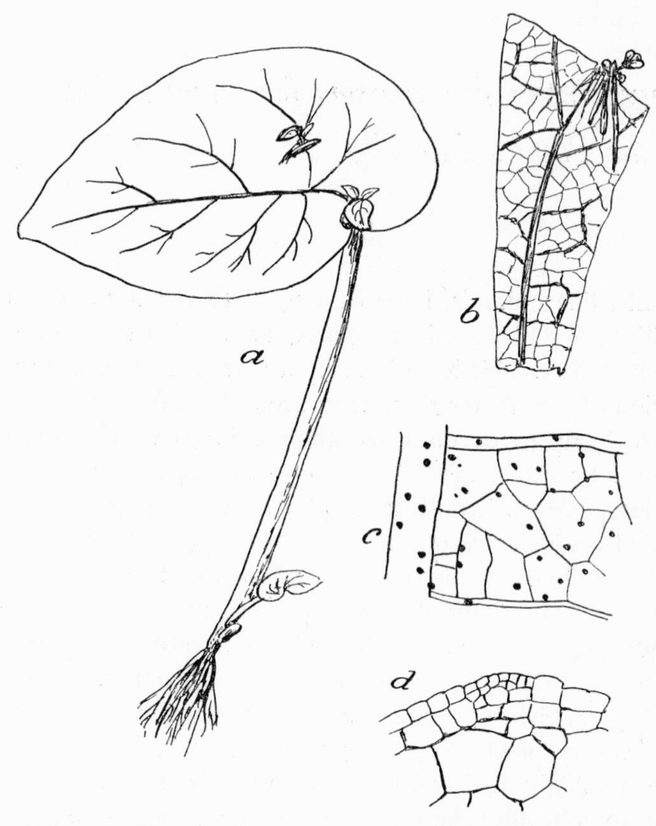

Fig. 25. Begonia (*Begonia rex*): (*a*) leaf kept in water regenerates in three places — at the base of the petiole, at the base of the blade, and above a cut lateral leaf vein; (*b*) application of auxin paste may cause regeneration at the apical ends of a leaf segment; (*c*) leaf surface, magnified ten times, with dots marking the primordia of adventitious buds, due to the influence of "colombine"; (*d*) section through a part of the leaf where one epidermal cell has begun to divide and form a bud initial (b-d from Prévot).

themselves by forming roots, the producers of organic nitrogenous substances, especially amino acids. It is probable that the formation of the root is stimulated by specific substances produced by the leaf. The same thing can be observed with isolated segments of stems, which regenerate roots more readily than buds. Out of 410 species of plants of which isolated segments were tested, 67 regenerated only roots, 27 regenerated roots and buds and 11 only buds (Plett, 1911). In these cases buds were developed either from inner tissues, especially from the cambium (callus), or from surface tissues already differentiated into epidermis, which reverted to the embryonic state and formed new cells resulting in a vegetative apex (Fig. 25d).

Begonia rex has usually been the textbook example of such complete regeneration of a plant, for it readily reproduces vegetatively from isolated leaves or even from sections of leaves. Such sections form buds on the callus, and buds sometimes also develop from epidermal cells, situated near the larger veins (Fig. 25a). This may be because of the retention of the products of assimilation in the leaves, but also because the isolation of the leaf sections stops the incoming current of inhibitory substances, which originates in the apical meristem and other growing parts. This example seems to show that the formation of adventitious buds on intact *Begonia* leaves is normally inhibited; it therefore occurs only if the connection between the leaf and the apical meristem has been severed. The same method of correlation holds here as in the case of apical dominance.

Inhibitory substances spread through the veins and not through the epidermis or parenchyma. Inhibitors other than auxin must be involved, for if we apply auxin paste to the base of the leaf we do not prevent its regeneration of buds. Perhaps related to this fact may be the inability of young leaves to regenerate, by contrast to mature leaves.

It is interesting to note that regeneration is exhibited only by epidermal cells, not by mesophyll cells. Even small pieces of epidermis, if connected to the parenchyma, can develop buds. Evidently this process is under the control of translocated substances, since it can occur on any part of the leaf, not only at the base or near a severed vein but anywhere on the leaf surface, even between the fine network of venules that separate the larger veins. Thus practically every epidermal cell can give rise to a new plant. This is particularly true in some varieties of begonias like *Begonia phyllomaniaca*, which got its name from a "mania" for forming a miniature forest of buds on the surface of its leaves.

Prévot (1939) obtained such regenerations by watering the plants with "Colombine," a 10 percent dilution of pigeon feces fermented in water (Fig. 25c). "Colombine" promotes the formation of buds even on the intact plant, and presumably must contain some substances capable of counteracting inhibitors. The same effect was obtained by placing the leaves for a certain period in a space without oxygen; buds grew even on the upper sides of the leaves when surrounded by an atmosphere of pure nitrogen. Another way to obtain the same effect is by applying auxin paste to the tip of the leaf (Fig. 25b). All of this indicates that what is involved here are either stimulators of a hormonal nature, or else compounds of some other type which are formed under anaerobic conditions. The formation of buds and roots near a sectioned vein is therefore related to an accumulation of auxin, rather than to the blockade of the normal product of assimilation such as carbohydrates.

The hypocotyls of some embryonic plants are known to show a similar ability to regenerate the whole plant from epidermal cells. If from a young seedling of flax[1] we cut off both the cotyledons and all the buds, then one cell of the hypocotyl will presently start to multiply and will form a new bud. Since the formation of such a bud would have been

[1] *Linum usitatissimum*. (Ed.)

inhibited had the plant been growing normally, inhibitory influences must have been reaching it from the growing parts of the plant. However, the cotyledons alone, without the buds, are not enough to prevent the formation of adventitious buds on the hypocotyl. The inhibitory effect is again interrelated with auxin, since if auxin paste is applied to the hypocotyl stump, regeneration does not take place. (This can be compared with the inhibition of axillary buds on seedling pea or bean plants, when the stump of the main axis is covered with auxin paste.) Usually triiodobenzoic acid acts in the opposite way; its application to flax hypocotyls brings about an increase in the production of adventitious buds, even in the presence of the cotyledonary buds (Fig. 2d).

Both of the above examples, in which new buds capable of giving rise to a whole new plant regenerate from mere epidermal cells, show that regeneration is primarily caused by a disturbance of normal correlations. Normally the formation and growth of the primordia leading to such a replacement of parts are inhibited by chemical substances, of which auxin is the best known and which very easily functions as an inhibitor. Without this regulatory action the plant organism would soon become chaos.

According to Kořínek (1922) one can measure "correlative sensitivity" in the case of the correlations acting during regeneration, for the amount of regeneration depends on the quantity and quality of the plastic materials involved.

It was shown above (p. 104) that when in *Caulerpa prolifera* the living substance was transferred into the rhizoids, only simple fingerlike shoots could be formed, in no apparent order (Fig. 7b). The same disorder is observed during the replacement of lost parts of trees. For example, after the trunk and the crown of a tree have been cut off, the cambium of the stump will form a callus which, at least in the horse chestnut, will give rise to a great number of buds. Such buds show many anomalies, such as fasciations, irregular leaf arrangement, or abnormal leaf size, all of which are signs of growth disorders; in fact most of these shoots die. Evidently the normal regulators, especially auxins, are missing. This is apparent also from other experiments on *Phaseolus multiflorus* seedlings, in which a lack of auxin brings about fasciations of the lateral buds that grow out after the removal of the main stem. Such fasciations become even more frequent in this experimental object if the level of auxin is lowered further, as with ultraviolet or X-rays.[2]

The fact that these buds develop so actively, in contrast to the weak regeneration of buds on leaves, is owing to the dominant action of the root system with its active amino acid metabolism. But even here a certain integration is reached, for when some of the many shoots become established they act as inhibitors for the others and eventually develop into new branches. The whole tree can thus be regenerated. This results, however, not only from the formation of new buds, but also from the activation of buds that have been dormant on the plant since the embryonic stage.

A common example of such regeneration is presented by cuttings from the black poplar, *Populus nigra*. Sections of branches from which all buds, active or dormant, have been removed, will form in a humid atmosphere a ringlike callus thickening at both ends just over the sectioned cambial cells. If this ring is removed the formation of callus by the neighboring cells of xylem and phloem is increased. At the apical end of the section normal buds will form on the callus, unless auxin paste is applied to the section at the beginning of the experiment. Thus auxin greatly favors the formation of the callus, but inhibits the regeneration of buds. This very important phenomenon was first discovered by Němec (1930) during his experiments with the roots of *Cichorium intybus*,

[2] Recent work suggests that these phenomena may not be due to simple "lack of auxin," but rather to a lowering of the ratio between auxin and a kinin. See, e.g., Skoog and Miller on tissue cultures (1958); Wickson and Thimann on axillary buds (1958, 1960).

which also form buds readily on their upper sections, from a relatively small callus. Němec divided the surface of the root section in half, separating the two halves by a tiny groove. On one half he applied a medium to which was added cultures of various kinds of bacteria, especially *Pseudomonas tumefaciens*, while the other half was left bare or provided with the same type of soil, but without bacteria. This control half produced a small callus with a great many leafy buds; the treated half gave rise to no buds, but its callus became very large and formed roots. These experiments were the first to show that the effects of auxin on the formation of buds were opposite to those on the formation of roots.

At the moment, auxin appears to be one of the most important prerequisites for the organization of root primordia. Auxin deficiency, on the other hand, causes an increase in the ability of the plant to form adventitious buds. As long ago as 1908, Němec showed that isolated roots of *Taraxacum officinale* very easily form new buds at their upper end. Even if only a small fragment of root is left in the soil, a new stalk soon grows from it and forms rosettes of leaves above the ground (Fig. 31d). Němec was interested in knowing how small a piece of root would suffice to regenerate the whole plant. He had discovered that long sections of root formed buds at the upper end and roots at the lower end; if, however, the sections were only about 1 mm long, and therefore very poor in nutrients and specific substances, they formed buds at both ends.

But the relationship is not as simple as it seems, since auxin alone does not always bring about the formation of roots; that is, there are many cases where after its application rooting does not occur. Chailakhian and Nekrasowa (1956) demonstrated that in some such cases roots may be obtained by the addition of thiamine to the auxin. There is a whole complex of conditions that must be present for roots to appear, and the same can be said for the regeneration of buds. These conditions are lacking in the cases where regeneration does not occur. Another example of this, namely incompletely regenerating leaves, was cited above. Similarly, buds are never formed on corn seedlings which have lost the upper part of the stem at the node which bears the coleoptile. There is no regeneration of the main axis from the mesocotyl, and all the reserve substances of the seedling are wasted on the formation of roots. The pea seedling behaves in just the opposite way, since its plumule can be replaced several times by the growth of buds in the axils of the cotyledons. Even after repeated removal, axillary buds can still regenerate from the meristems in these axils (Fig. 16).

Relatively little is known about the substances that promote the formation of buds in the way that auxin acts on the formation of roots.[3] This problem can be studied better in tissue cultures, where the composition of the nutrient substrate can be closely controlled. Here the formation of buds and that of the mesophyll of leaves is promoted by adenine and related compounds. However, contrary to what happens with roots, such new formation is inhibited by various factors, as is shown by the following experiments. These bring out at the same time the relationships between leaves, roots, and buds. If an isolated segment of the stem of Scrophularia, with one node and one pair of leaves, is cultured in water, then, after one of the leaves has been removed, the roots will grow more actively on the side of the remaining leaf, and the bud on the side from which the leaf has been removed. Optimum conditions for these experiments seem to be low light and high humidity. They are successful even with fully grown leaves, which if left on the plant would have produced only reserve substances for the formation of tubers. However, from stems at the end of their vegetative period neither roots nor buds can be formed, but only a callus, which appears on the lower section. The callus is thus the "last effort" of the

[3] See note 2, this chapter.

plant toward integration, since it is formed by the leaf products when inhibitions are strong enough to prevent the growth of buds.

Isolated cotyledons of the pea, which usually regenerate roots very easily, under some conditions form only a voluminous callus which exhausts their reserves and they die. On the other hand, there are numerous examples among the squash (Cucurbitaceae) family, where the cotyledons regenerate both buds and roots and thus reform a whole plant. The formation of the callus itself can be prevented by applying maleic hydrazide to the experimental plants. Sections of poplar stem spread with maleic hydrazide paste maintain a completely flat surface without a callus. Meanwhile, adventitious buds can develop from the scars left by the removal of lateral buds, showing that the inhibitory influence of the maleic hydrazide does not reach these organs. Regeneration can thus be controlled by a number of chemical substances, though the influence of external factors must not be overlooked.

Some phenomena seem to oppose the idea of integration in the plant. For instance, organs often regenerate from cells which are inappropriate to them. *Bryopsis muscosa*, another Siphonaceous alga like *Caulerpa prolifera*, has a feathery stem and a branched, colorless rhizoid. If the stem is cut off a new one will grow from the cut surface, and if the rhizoid is removed another rhizoid grows in its place. But if a segment of the stem be placed horizontally, keeping both ends in the light, then on both ends only stems will form. Here light has caused the growth of a light-dependent organ at a site which is normally in the dark and carries rhizoids. Similar behavior occurs in the roots of the dandelion (*Taraxacum officinale*) or the carrot (*Daucus carota*); if these are planted in the soil upside down, so that the normally upper end is in the dark, then on the lower end, now on top, we witness the formation of buds rather than roots. Light thus opposes the influence of auxin and also promotes the regeneration of buds, even if it means acting against gravity and against those internal correlative conditions which result in polarity. The same disturbance of polarity can be observed on aerial organs which normally form roots only on their lower end; if they are surrounded by humid soil or if by damaging the surface layers (for example, on willow branches) we permit a direct contact with water then they can form roots at the upper end also or at any other site. Thus regeneration can be controlled by nonchemical influences which can decide, for example, which of a group of existing primordia will develop most actively. This is true also for those root primordia and buds which develop on the convex side of curved stems or roots.

Our present knowledge of the nature of regeneration leads us to view these phenomena materialistically. Both in regeneration and in reproduction, that is to say in the development of primordia originally present but now removed, one may say that in general the missing parts of a plant are regenerated at a different site from where the scar was left. This behavior is different from regeneration in animals, which occurs on the actual site of the wound, as in the case of the head or tail of a planarian or the tail of a lizard. This might be explained by the typical organization of plants in a great number of segments or metamers, which make possible great numbers of primordia and thus an easier replacement.

In plants a regeneration comparable to that of animals occurs only on vegetative tips. Němec (1905) gave a very thorough exposition of the regeneration of the root tip, ascribing an important influence to specific substances that are independent of the quantity of food, both for regeneration and for growth correlations in general. If a piece of the root tip, 0.75 to 1 mm long, is isolated, a low callus appears on its cut surface (Fig. 26a), then a new vegetative tip forms which later produces a root cap and ultimately the rest of the root body (Fig. 26b). To obtain such restitution it is necessary to sever the connections between the vegetative tip and the rest of the apical meristem, and for this it is sufficient to cut at least halfway through the so-called procambium (Fig. 26c).

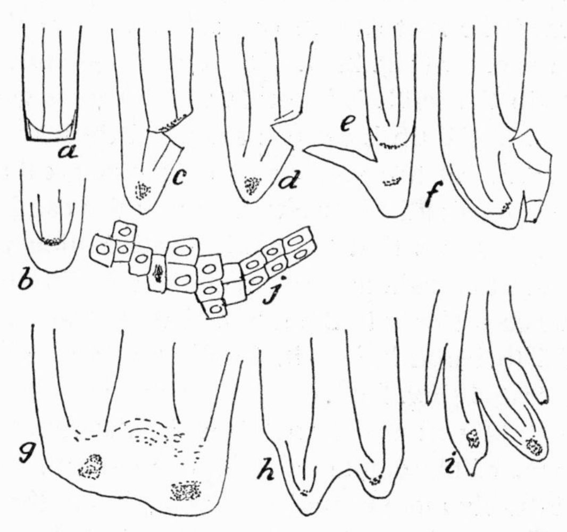

Fig. 26. Broad bean (*Vicia faba*), regeneration of the root tip, after Němec: (*a*) when root is decapitated less than 1 mm from the tip a continuous callus is formed on the wound; (*b*) after three days a newly formed meristem and root cap appear; (*c*) root cut deep into the plerome forms a new apical meristem in two days; (*d*) if the cut is not deep enough (inner plerome has remained intact) no new meristem appears; (*e*) root cut obliquely from above, after five days, with regenerated cap and meristem; (*f*) after two cuts at the same level, regeneration above the deeper cut (left); (*g*) root decapitated 1 mm from the tip regenerates two tips; (*h*) same two days later; (*i*) three longitudinal cuts lead to the regeneration of two new tips in two days; (*j*) meristem of *i* magnified, four days after operation.

But if the cut is deep enough for the wound not to heal, an intercalary regeneration results. Similar sectioning leads also to intercalary regeneration in animals like the hydra, the planarian (Fig. 31g), the lizard, and the tadpole. Just as the tail of a lizard can regenerate repeatedly, so the root of *Vicia faba* can respond several times to such sectioning; however, the cuts have to be made on opposite sides of the root, for if they are on the same side only one regenerate appears. Also the cuts must be equally deep, otherwise only one new initial layer of apical meristem will regenerate (Fig. 26f). The initial layer consists of dividing cells, which give off new cells toward the root cap on one side and toward the root body on the other (Fig. 26g). If the cut is made further away from the tip the result is like that which occurs when more than 1 mm of the tip is cut off: a normal tip is not restored but rather several tips are formed on the ring callus. If the sections are longitudinal, two to four new tips can be formed (Fig. 26i). It seems that the meristematic region when disturbed tends to regain its former state of equilibrium. Actual restitution occurs only in the terminal part of the root which is cut 0.75 to 1 mm in length. In all other cases several tips are regenerated (Fig. 26g).

An attempt was made to follow in the same way the replacement of the vegetative apex of the stem. Conditions are considerably more complicated here, due to the closeness of leaf and lateral bud primordia to the apical meristem. This situation has made such experiments more or less a failure, for the damaged apex was not reformed on the site of the wound and one could therefore never consider it as a restitution. The same difficulties were encountered in the case of the noncellular alga *Caulerpa prolifera*, in which the analogues of vegetative apices are somewhat more accessible, being formed by whitish spots on the rhizoids and rootstock. "Leaf" primordia are also at first only white dots, which later, when the blade enlarges, elongate into a narrow white strip containing cytoplasm without chloroplasts, starch grains, or vacuolar contents. All structure appears to be absent, for as soon as they are touched these zones turn yellow and on their border new apices arise.

It is interesting to follow the regeneration of the surface area of an assimilator when its vegetative tip has been damaged. It is enough to leave the plant out of the water for a little while for the tip to stop growing completely. Regeneration will then take place at the site of the wound, but the new blade will grow at an angle of 90° from the original one (Fig. 6f). This may be analogous to the crosswise arrangement of the first leaf primordia below the apex of plants with opposite leaves. Such an arrangement is also most simply explained by inhibitory influences at work among the youngest leaf primordia. Similarly with *Caulerpa prolifera*, when the vegetative tip is greatly damaged two new apices will grow laterally and the "leaf" will develop into two long blades. This is reminiscent of the forked branching which occurs in the fern *Phyllitis scolopendrium* (*Scolopendrium vulgare*) when the apical meristem has been damaged. Once more we see a similarity between the processes carried out by the unicellular alga and those of higher plants. The differentiation resulting in the external shape of the alga must be caused by correlative substances originating in the outer layer of cytoplasm, or "ectoplasm," attached to the cell wall and transported by the protoplasmic circulation to all parts of the plant.

Caulerpa prolifera is also capable of a healing process, although a much simpler one than that found in higher plants; in higher plants a callus or a periderm is usually produced by a meristematic ring, the phellogen, which on the outside forms the protective cork. Instead of this, the alga closes the wound with a vacuolar coagulate, and this stopper prevents the content of the cell from escaping. New membranes, formed of hemicellulose and pectins, are then deposited on the inside. This phenomenon is important for the propagation of the plant, since each isolated piece can heal its wounds and form rhizoids, "leaves," and finally a rootstock, thus growing into a whole new plant. In many cases something similar to heteromorphosis occurs, perhaps a rhizoid growing instead of a rootstock, or the tip of the rhizome developing into a "leaf." This is controlled again by internal conditions, in the first case by the loss of rhizoids and in the second case by the loss of leaves from the whole plant. Therefore, as has been said so many times, regeneration "restores what has been lost."

This statement, however, does not explain the nature of the process of regeneration. Such an explanation would require a thorough analysis of the metabolic changes that take place in the plant when correlations are altered. It has been suggested that during healing the wounded cells produce a traumatic acid which spreads to other tissues, where it causes the cells to become "embryonic," resulting in the formation of phellogen and cork.

On the other hand, there are plants which do not regenerate so easily, or which do not even form a callus after they have been damaged. The celandine, for instance, probably does not possess the tissues necessary for regeneration. Its buds are formed only in the axils of the leaves, while its roots form only on axillary buds and from the beginning become differentiated into tubers. These tubers are normally spherical but can be forced to elongate by amputating the basal ones and growing them in water. Normal absorption roots, though few in number, do appear at the base of buds situated on very short internodes and preparing for the next vegetative period; tubers develop above these later. On very small tubers which have become independent from the main tuber and have formed buds for the next vegetative period, the short internodes between the scales may elongate, the amount of elongation depending on the depth at which they are situated in the soil. On such an elongated tuber or rhizome, roots may form on the internodes, and these roots can be changed into tubers by treating the internodes with auxin (Fig. 8a and b). This does not mean, however, that auxin is the specific substance capable of producing tubers or of transforming absorption roots into storage roots, but rather, as in the other cases, that auxin increases local inhibitions, thus favoring the formation of tubers over that of roots.

During the elongation of the stem, inhibitions continue to

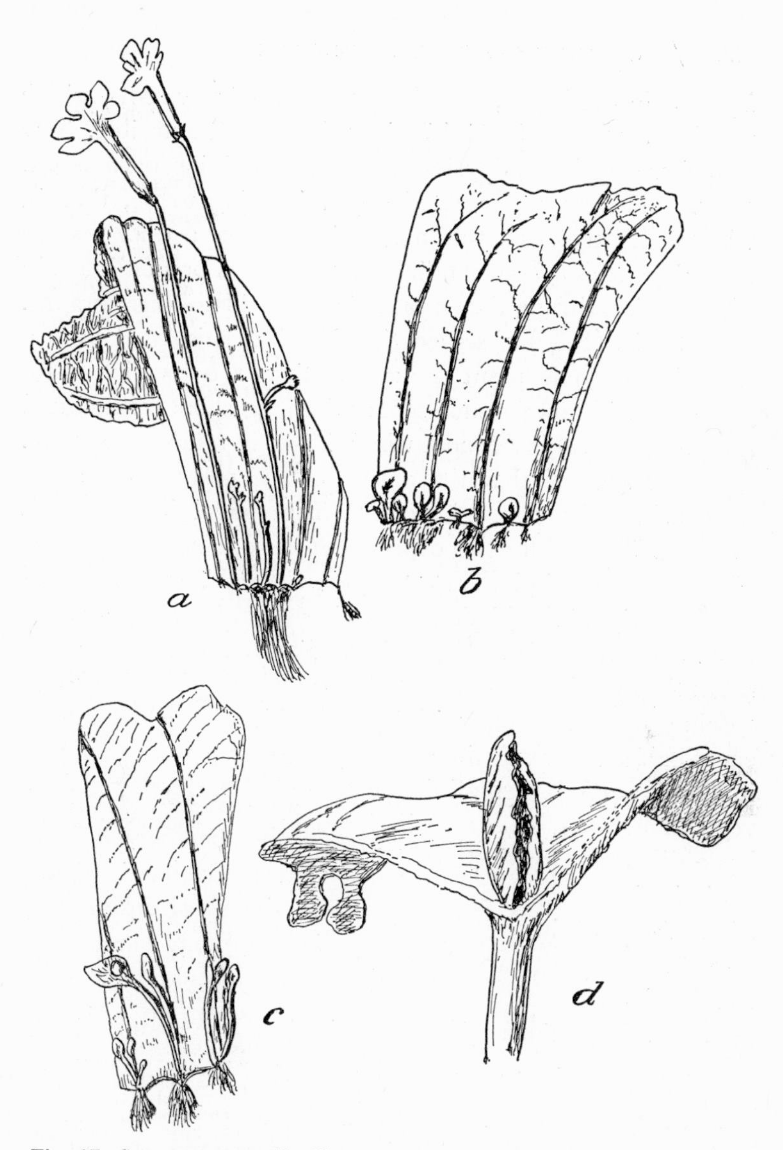

Fig. 27. *Streptocarpus wendlandii*, segments from different areas of the same large blade, corresponding to one cotyledon: (*a*) flowering shoots regenerate from the basal area; (*b*) leafy stems regenerate from the apical area; (*c*) transitory forms with leaves and flower buds regenerate from the central area; (*d*) winglike outgrowth, following a cut 5 mm long and 0.5 mm deep into the basal meristem of the otherwise intact cotyledon (after Němec).

increase in the leaves, and the formation of new roots, which requires a low level of inhibitions, becomes impossible. New branches sometimes develop from the remaining parts of old roots. What the regenerated organ is going to be, that is, whether it becomes a thin absorption root or a thick reserve root, depends in the celandine on substances predominant in the organism during the embryonic formation of these organs. Thus Sachs (1880) observed that if a leaf is isolated from a flowering Begonia plant the adventitious shoot which it regenerates will flower much sooner than the shoot regenerated from a comparable Begonia which is not in a flowering state.

Němec showed that the nature of the regenerated shoot depends on the part of the leaf from which the shoot is growing. He used for this experiment the one large cotyledon of the ornamental plant *Streptocarpus wendlandii*. Sections taken from the tip of the blade gave only vegetative shoots, while those taken at the base almost immediately produced flowers. From the middle part of the blade were obtained intermediate specimens, which began with leaves and later developed flowers (Fig. 27a, b, c). If the regenerating blades were wrapped in aluminum foil, then all the parts of the leaf developed only vegetative shoots. Thus each individual part of the same organ has a different growth potential, modifiable by light.

Similarly, we can divide the stem of *Scrophularia nodosa* or *Circaea intermedia* into sections, each with one pair of leaves, and even if we culture them all under the same conditions the buds will develop differently, depending on their point of origin on the stem. In sections taken from the base the axillary buds will form runners and tubers; those taken from near the apex will give flowers (whether the donor stem was flowering or had just finished doing so); sections taken from the middle part will produce leafy shoots which may flower later (Fig. 28a–d). If, however, the leaves are cut off, or even wrapped in foil, then all buds will develop only into weak shoots. Later in the summer these conditions change, all the sections then giving rise only to runners or tubers. Finally, at the end of the vegetative period, the aging leaves produce substances which permit only the growth of small, weak, leafy branches.

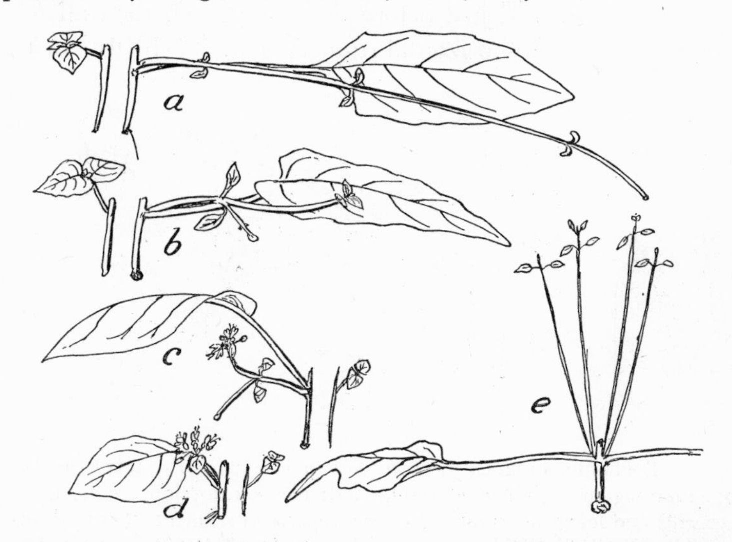

Fig. 28. Enchanter's nightshade (*Circaea intermedia*), segments taken from different levels of a leafy stem, with originally insignificant buds, divided longitudinally in two halves, one of which is defoliated: (*a*) runner forms on the basal segment; (*b* and *c*) transitory forms appear on the middle segments; (*d*) upper segment forms flowering shoot on the side where leaves remain; defoliated half forms only weak, leafy shoots; (*e*) basal segment treated with gibberellin paste forms normal shoot (cf. *a*, without gibberellin).

Lysenko (1952, 1954) observed similar phenomena with stems of the soybean; basal sections gave rise to leafy shoots, apical sections to flowers. He concluded, therefore, that basal sections are at a younger stage than the apical sections.

Tubers of *Circaea intermedia*, when divided transversely into sections, form branches which differ from section to section. Sections closest to the base of the tuber produce small horizontal runners which eventually penetrate into the substrate and thicken into tubers; the uppermost sections produce vertical leafy shoots; and middle sections give rise to intermediate forms, that is, to obliquely growing shoots with smaller leaves (Fig. 29a–d). If this same experiment is performed at the end of the rest period, then all the sections will form vertical leafy shoots; if performed just before the rest period, then all the sections give rise to horizontal runners (Fig. 29e–h). It can be

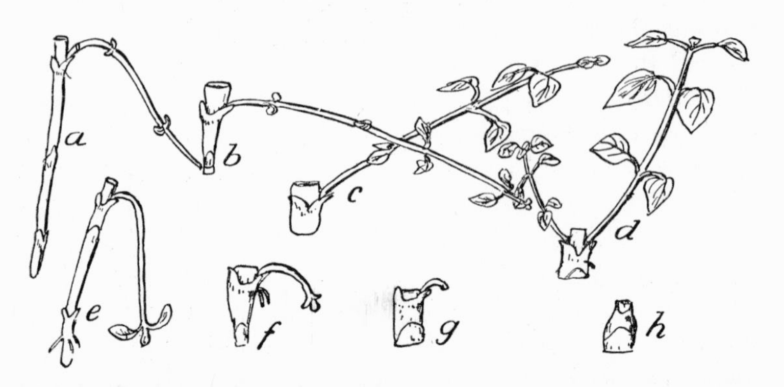

Fig. 29. Enchanter's nightshade (*Circaea intermedia*): (*a* to *d*) tuber divided into four segments at end of rest period; runners develop on the basal segments and leafy shoots on one of the uppermost segments; (*e* to *h*) resting tuber similarly divided; runners develop in all but the most apical segment.

concluded that there must be a constant turnover of the substances responsible for new growth and differentiation. These phenomena have not yet been clarified biochemically or biophysically. When the plant is well integrated all the organs are synchronized and the regional differences disappear.

Thus far discussion has been confined mainly to buds and roots, since these are the organs most frequently formed after the loss or exhaustion of certain organs. Leaves, on the other hand, do not re-form independently of the stem, though young specimens of *Cyclamen persicum* present a rare exception to this rule. If we cut off the blades of the primary leaves which develop immediately after the cotyledons, then there will grow out, directly below the cut surface, two small winglike blades which may later fuse into one. If the blade is not removed but merely rendered inactive by plastering it over or coating it with collodion, similar outgrowths will result. The other leaves, however, are not capable of such regeneration.

In other plants, we have noticed that if all the leaves, and all the buds which could replace the leaves, are removed, then the stem itself becomes much greener and the subepidermal cells elongate vertically toward the surface, simulating the palisade parenchyma of leaves. In this way there are obtained, artificially, assimilatory stems like those in plants with degenerated leaves, such as *Sarothamnus scoparius*. In the pea seedling, if all the buds are cut out, the cotyledons become green and show the same transformation if kept in the light. In general it seems that if the primordia of storage organs are removed then the storage products flow into other tissues, which alter their function and correspondingly also alter their structure. Thus sections of ordinary stems, petioles, or roots can be changed into tubers; their cells enlarge and multiply, especially in the region of the pith and cortex, and fill up with starch. The conducting tissues also multiply, while supporting tissues such as the collenchyma are suppressed. Sometimes such major anatomical changes are not permanent because the newly formed tubers are not able to form vegetative tips or buds. This shows again in nature a determinism independent of finality.

In plants like the sunflower, which are incapable of replacing lost reserve organs, and whose tissues are unable to undergo such major changes as those mentioned above, cutting out the buds causes the axillary meristems to grow into large tumorlike formations. Isolated tubers of dahlia or of celandine, which do not carry lateral buds, do not suffer exhaustion of their reserves, and thus remain alive for a long time. On the rounded end of the tuber of the celandine a callus tumor develops. This consists of parenchyma cells, some of which transform into tracheids, but there is no differentiation of roots. What has remained of the apical meristem of the original root probably started to divide but was somehow inactivated by the reserve contents of the tuber, so that the division was not regulated. It would follow that the radial vascular bundle of the tuber does not influence the differentiation of tissues inside this unusual callus. In other cases it has been shown that the appearance of the radial vascular bundles in the tip of the root depends, at least in part, on the already differentiated bundle to which they are joined. The reserve content of the tuber also inhibits its branching, so that it will form lateral branches after decapitation only while it is young and thin. Auxin especially favors such branching, just as it promotes the branching of normal absorption roots.

The apical callus of the celandine tuber which is unable to regenerate buds or indeed any callus formation in general, is an expression of the plant's tendency to retain its integrity even though for the time being the differentiation of new organs is impossible. Callus may form anywhere the flow of translocating material is stopped, as may be seen on trees which have been ringed all the way down to the cambium. As long ago as 1727 Hales showed by such ringing that water and minerals ascend from the roots to the top through the xylem, and it has long been known that the products of assimilation travel down through the phloem. Ringing stops this flow and the block leads to the formation of a callus above the ring.

A common method of studying the development of callus is by the use of sections of poplar branches suspended freely in a humid atmosphere. These form an extensive callus, composed of large white cells, on both ends. The callus on the lower side, however, is the larger, even when the branch is lying horizontally, a condition which must correspond to a block in the transport of auxin on the lower side. Roots show similar behavior, although in some plants the larger callus is found at the upper end (dandelion) while in others (alfalfa) it occurs at the lower end. This may be because the flow of auxin here does not have such a definite direction as in aerial parts which bear leaves containing the auxin precursors.

The flow of auxin is probably also responsible for the differentiation of conducting tissues, especially the vessels of the xylem. Jost showed in 1891 that if growing leaves are grafted on to very young plants, vessels will form in unusual places. He finds that new and old veins become joined through a connecting vein which grows out from the top of the old vein. The same thing happens in stems; parenchyma cells are transformed under the influence of the flow of auxin into conducting elements and so produce connections between cut veins.

It was known even before the discovery of growth substances that there is a close relationship between the unfolding of young leaves from buds and the formation of seasonal rings in the trunk of trees. If the leaves are lost through late frost or insect attack, at the beginning of the formation of summer wood, then a double ring may be formed in that year. It has been shown directly that the activity of a cambium which has gone into a resting state can be restored by application of auxin paste. However, the exact causes of further differentiation of the cambium are still unknown, for differentiation into specific tissues cannot yet be assigned to specific growth substances. Supporting tissues have been studied a great deal in the hope of discovering if they could be strengthened or altered under the influence of pressure or tension. The results show that internal conditions present in the earliest stages of development, long before any external factors can be applied, are of primary importance. There are probably special correlations among individual cells which are at first totipotent but later differentiate according to their position relative to other cells (Razdorsky, 1955).

For example, the growing fruit of the cucumber determines the development of mechanical tissues in its stalk, but outside mechanical pressures applied to the stalk in the absence of a developing fruit are quite without effect. Experiments which show this very clearly depend on placing an organ with almost undifferentiated conductive tissues in the main pathway of assimilatory products traveling from the leaves to the roots. Thus a potato tuber will stay alive much longer and form many more vessels if it has a stalk on one end and a root on the other. A similar phenomenon is observed if part of a pea cotyledon is cut off of a piece and to the cut surface the tip of the main stem is grafted. This tip may then fuse with the section and grow out into a stem, providing care was taken to cut out all the buds on the rest of the plant. Even such an ephemeral organ as a cotyledon, which usually shrinks as soon

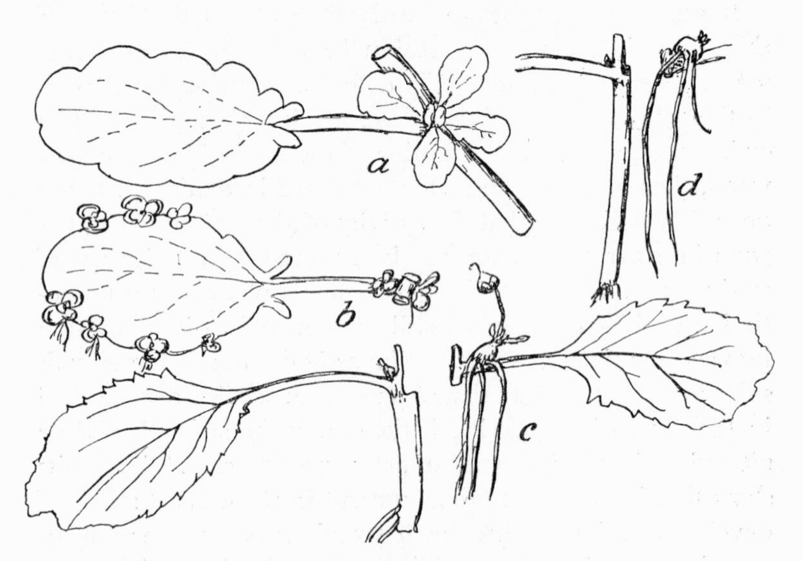

Fig. 30. *Bryophyllum crenatum:* (*a*) axillary shoot develops on a segment with a long internode, but marginal buds remain dormant; (*b*) segment with very short internode develops marginal buds. *Scrophularia nodosa:* (*c*) axillary bud grows less on segment with long internode than on one with only a short piece of the axis, where bud swells in tuberlike fashion and forms a capsule; (*d*) segment with small piece of axis develops large tuber while opposite bud, on segment with long piece of axis, scarcely enlarges.

as its reserve content is used up, will in this case form thick vascular bundles and maintain the connection between the root and the grafted stem.

An interesting example in terms of the relation between regeneration and restitution comes up when the plant is wounded without removing any tissue. Němec (1911) described experiments showing regeneration of the base of young leaves of *Streptocarpus wendlandii*, the meristems of which were cut in half along a longitudinal line. The blade developed large overlapping lobes on both sides (Fig. 27d). Similarly in Snow's (1933) experiments, a longitudinally slit stem was shown to reconstitute all the tissues on each side of the cut, so that two axes arose next to one another although nothing had been lost.

It follows that there is not only a longitudinal but also a transversal and circular (tangential) flow of substances which tend to preserve the overall shape of the plant externally and internally. The longitudinal flow is the most important because most plants develop in that way, though it remains so even in tissues where width exceeds their length, as in the aerial tuber of Kohlrabi, which contains a network of vessels in all directions.

The significance of the axis is shown also in Fig. 30. A long section of a stem of Bryophyllum forms one large axillary shoot and no marginal buds, while a short section, in which movements of products are blocked, has strongly developed marginal buds and the axillary buds are inhibited.

CHAPTER V

Polarity as an Expression of Integration

As we have seen (Chapter IV), many plants show a clear-cut polarity in that isolated sections of stem, maintained under favorable conditions, regenerate buds at the apical end and roots at the basal end. This behavior is one expression of the polarity which governs the normal development of the whole plant, and is a natural consequence of the adaptation of the plant to two different food sources. On the one hand a plant obtains water and mineral salts from the soil through the roots, and on the other hand it receives light energy and CO_2 from the air through the leaves. Recent conjectures on the origin of life, and how the present forms of autotrophic green plants came about, lead to the concept that these two oppositely moving modes of supply are deeply rooted in the protoplasm of plant cells. For this reason, every larger section of a plant contains a basal and an apical pole oriented in the way the plant was oriented in the soil. In the root the situation is the reverse of that of the stem, since its apex points downward.

If the organism consists only of a single cell, outside influences will create a polarity in it. A good example is furnished by the egg (zygote) of an alga of the Fucus family. In this cell chloroplasts are at first distributed evenly throughout the cytoplasm. However, on exposure to a light source from one side, all the chloroplasts accumulate on the lighted side. The colorless part of the egg later becomes separated from this green part, by a division which takes place perpendicular to the direction of the light, and it then forms rhizoids (Fig. 13a). A similar polarity can also be brought about through the action of gravity or by centrifugation, when the lower part or the centrifugal side, respectively, form rhizoids. In an electric field, rhizoids form at the positive pole (Fig. 13d). It is notable that the application of auxin on one side of the cell brings about the formation of rhizoids on that side. Everything therefore suggests that auxin, which here stimulates the formation of rhizoids as it does that of roots in higher plants, participates in an important way in the establishment of the polarity.

The first scientific[1] explanation of polarity was given by Němec (1930) in his previously mentioned experiments with the roots of *Cichorium intybus*. These roots were cut into sections, and on the basal cut surface a bacterial culture producing auxin was applied. Normally, vegetative buds would have formed at this pole. However, when auxin is present, there are formed a callus and roots. The localization of these roots is undoubtedly the main indication of the polarity.

A common example of polarity is presented by sections of Salix (willow) branches suspended in a humid atmosphere horizontally, so that the force of gravity acts equally on both ends. After a time, roots always form at the basal end and buds at the apical end. This localization is more precise if the branch is held in a normal, upright position, and less so if upside down, for then gravity counteracts the inner influences. Auxin is apparently partly displaced in response to gravity and tends to flow from the basal end (pointing upward) to the apical end (pointing downward). On horizontal branches, roots form the more actively toward the lower end,

[1] The word used is "materialistic." (Ed.)

while buds develop more toward the upper end. Certain correlations are soon established between the various primordia, larger ones inhibiting smaller ones, with the result that the primordia situated halfway between the two poles become completely inhibited. If the section is cut in half, two new poles appear, for the transportation of inhibitory influences has been severed. Polarity is therefore a correlative matter.

It is for this reason that even in higher plants correlations originate in the single cell from which the plant originates. In a normally fertilized egg, influences comparable to those seen in algae bring about a permanent differentiation of two poles, one of which subsequently produces the radicle and the other the plumule. This differentiation is preserved throughout the life of the plant. It is probably controlled by the structure of the protoplasm itself, and it has even been ascribed to the asymmetric structure of the constituent proteins, with their NH_2 and COOH poles (Sabinin, 1957).

Vöchting (1906) tried to change this polarity by rotating a plant egg cell on a clinostat so as to balance out the effect of gravity. He did not succeed, however, for external factors do not act in the same way on higher plants as they do on lower plants. The establishment of polarity seems to occur even earlier in the development of higher plants and is more or less irreversible. A great deal of effort has gone into trying to reverse polarity. For decades young trees were kept planted upside down, with buds and shoots cut off on the now lower part, while root primordia were removed from the now upper part and only shoots which appeared there were left on. Yet, after all this, when sections of the stem were isolated they still possessed the original polarity.

On the other hand, polarity can be concealed by local application of auxin on various sites of the stem. In the very first experiments with auxin it was shown that auxin causes the outgrowth of roots on stems of bean or pea seedlings, although the normal roots were still present. An intact stem of a tomato or of *Coleus* coated with a paste of auxin in lanolin forms a dense feltlike covering of roots. Before the discovery of this substance, an apparent reversal of polarity had been obtained by increasing the supply of water at the apical end of that classical object a branch of willow, whereupon roots started to develop there. It appears, however, that the formation of roots is related to that of bud scales. In pea seedlings, roots develop very easily along the whole length of the two lowest internodes, which are terminated by primary bud scales, but not on the subsequent internodes which bear true leaves. If, however, the stem is cut off above the cotyledons, keeping both lower internodes in place, roots form only very close to the basal cut surface of the stem.[2] Green leaves, which produce auxin, also probably cause it to accumulate immediately above the cut surface. In intact plants, auxin precursors, originating in the cotyledons, spread upwards and are there transformed into auxin. This auxin, when activated, brings about the formation of roots on the lower internodes.

The influence of bud scales is apparent from the fact that roots are usually formed on the nodes of runners, where scales are borne, and also on underground tubers, though with the well-known exception of the potato tuber. This exception can be explained by a strong inhibitory influence which interferes with the activity of growth factors, so that root primordia form without apparent polarity only on the sprouts near the scales and on axillary buds. However, under the influence of auxin even tuber tissues can form roots.[3]

In some cases, as when, for example, the apex becomes transformed into another organ such as a tuber, the polarity of the stem appears to have been altered. Klebs (1903), who in his experiments tried to interfere with normal plant development, obtained many such results, an apex sometimes form-

[2] I.e., cutting off appears to accentuate the polarity. (Ed.)
[3] Very high auxin concentrations are required, however. (Ed.)

ing a rosette from which roots could develop. Vegetative plants of *Bryophyllum crenatum*, at the end of the summer vegetative period, form short internodes with abnormally small leaves and roots, so that in fact the original polarity has disappeared in them. But stem segments from plants grown during the long-day period maintain normal polarity. This same retention of the original polarity is found in the long hanging branches of the weeping willow. If an isolated section of such a branch, which normally hangs upside down, is placed in a moist atmosphere horizontally, it forms roots only at the basal end and buds at the apical end.

Some years ago an experiment was reported in which the polarity was altered by inverting the position of a shoot (Went, 1941). But this he explained on the grounds that the original cells kept their polarity, while new cells, especially those of the phloem, were adapted to the new position under the influence of gravity so that buds developed on the original basal pole and roots at the apical pole. Even here it is apparently through the phloem that the correlative influences, which are responsible for polarity, are transported.

If a willow branch is carefully ringed, buds will grow from the scar below the ring and roots from the scar above the ring (Fig. 31a). The xylem, therefore, is not involved in the transport of these factors. The formation of roots above the ring is due to the accumulation of auxin at this point, and the growth of buds is made possible because the auxin (which would have inhibited them) has been transported to the opposite pole and is not replaced because of the ring. Fischnich (1939) ringed willow branches near one of their extremities, making only small rings. The difference in auxin level then became very small, and both ends produced buds (Fig. 31b). This is again evidence that the lack of auxin stimulates the formation of buds. The same thing was shown by Němec in his experiments with the roots of dandelion (*Taraxacum officinale*) which also regenerated buds on both the upper and lower cut surfaces, if the section was only one millimeter thick. If large pieces of roots were used, buds regenerated only on the basal end, and roots at the apical end (Fig. 31d).

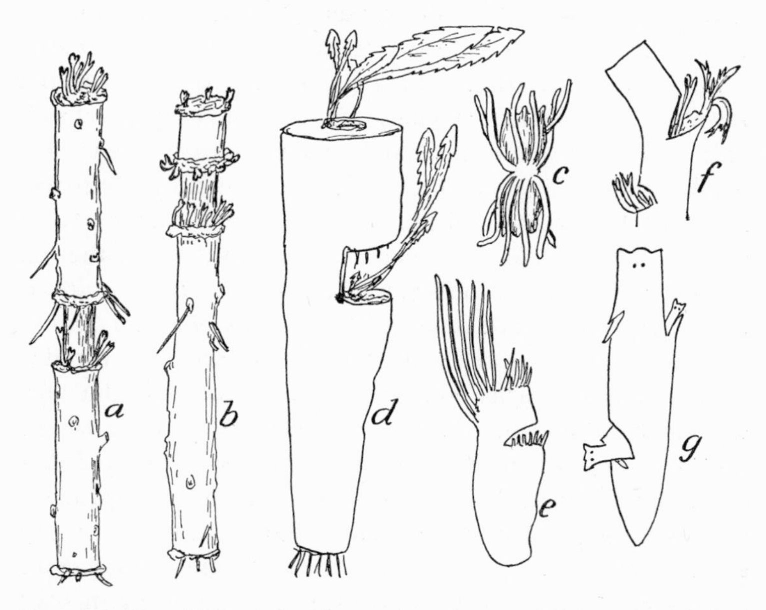

Fig. 31. Willow (*Salix alba*): (*a*) segment ringed in the middle forms buds at apical pole and roots at basal pole; (*b*) segment ringed close to apical end forms buds both above and below the ring (after Fischnich). Polyp (*Tubularia mesembryanthemum*): (*c*) polyp, sectioned immediately below the head, forms on the lower side tentacles and proboscis instead of rhizoids (after Loeb). Dandelion (*Taraxacum officinale*): (*d*) root cut laterally forms buds at the basal and roots at the apical end. Anemone (*Cerianthus membranaceus*): (*e*) anemone regenerates tentacles below a transverse cut, while above it the growth of tentacles is inhibited (after Loeb). Horse chestnut (*Aesculus hippocastanum*): (*f*) buds developing on apical surface of transverse sections (after Němec). Flat worm (Planaria): (*g*) head developing on apical section and tail on basal (after Morgan).

These phenomena are comparable to some encountered in animals. Large sections of the stem of the polyp *Tubularia mesembryanthemum* regenerate a head, with tentacles and a proboscis, at the oral pole and rhizoids at the opposite (aboral) pole, while if the section is performed just under the crown of tentacles, above the stem, the aboral pole develops a new "head" (Fig. 31c). In this case, however, we are not dealing with a true regeneration but rather a modification, in which the disconnected parts of the section transform rather quickly into folds and tentacles. Similar experiments were carried out by Morgan (1907) on Planaria. When a planarian was divided transversely into two halves, the upper half regenerated a tail and the lower half a head. If, however, the cut was made very close to the head, the head portion regenerated another head, while if the cut was close to the tail, the tail portion regenerated another tail. The polarity shown by these experiments is at least analogous to that in plants.

The root-forming capacity of stems is not uniformly distributed along their length. An isolated stem of Scrophularia, with its basal end in water, forms many roots. If, however, such a stem is divided into several sections, each consisting of one internode and one pair of leaves, then the most basal section forms no roots at all, the next one up only a few, while the middle and upper sections produce a great number of roots. Thus a polarity exists in the stem as a whole, permitting roots to develop only at its base; but this base when isolated is incapable of forming roots. In other words, the base of the stem is a zone of attraction for the root-forming activity of the whole plant; in this phenomenon resides the correlative nature of polarity. It is probable that auxin, which is being produced by all the leaves, accumulates at the base of the stem. Such an accumulation of auxin can be especially clearly deduced in axillary branches of *Bryophyllum crenatum*. These branches grow usually from the rosettes formed during short winter days and seem to narrow down at their point of insertion as if they were ringed. A similar phenomenon occurs whenever there is a sharp bend on a stem or a branch, for instance when a horizontal or an inverted stem straightens up under the influence of geotropism. Adventitious roots will appear at this point in spite of the inhibition exerted by the normal root system. If these adventitious roots are allowed to develop and then are planted in soil, and if the stem then is cut off at its original base, isolating it from its original root system, this base will still form roots even if it is pointing upward.

Such aerial roots are probably not just a useless expression of the polarity inherent in the plant, for Kursanov (1952) showed that they, along with the roots in the soil, contribute to the metabolism of nitrogen. Their formation demonstrates that polarity cannot be obliterated by external factors such as gravity, light, or lack of humidity, or even by internal factors like the inhibitory effect of the roots developing on the bend of a stem planted in soil.

Leaves, on the other hand, are not so strongly influenced by polarity. In the above experiment the oldest leaves, closest to the base, which on the intact plant were starting to turn yellow, become green again, while the younger leaves, which on the intact plant were situated near the tip and, in this experiment, above the bend, become dry and yellow. The conditions of polarity are therefore essentially reversed here; substances have been exchanged between the leaves of different ages, so that the oldest, now activated through the influence of the roots, remove not only water but also specific substances (especially proteins) from the younger ones.

To pursue this study further the following experiment was performed. One leaf, of the lowest pair of leaves on the stem, was isolated from the influence of the rest of the stem by carving out an oblique groove through the stem above it, while the opposite leaf was left in normal circulation. The leaf below the groove regreened even if it were already yellow, and a bud developed in its axil. Moreover there were no traces of marginal buds. The opposite leaf, on the other hand, turned yellow, its marginal buds enlarged, and the axillary bud remained completely dormant (Fig. 32). In this way the lower leaves of the ordinary axillary rosettes which develop later in the winter from the dry top parts of the plant also turn yellow, while only the supporting leaves stay alive. The lower yellow leaves, in spite of their small surface, form border buds (marginals), on which in turn new marginals grow, while at the base of the rosettes aerial roots are formed. In the corresponding aerial roots of Ficus, Kursanov (1953) found a num-

Fig. 32. *Bryophyllum crenatum:* lowest node on isolated stem with pair of leaves; above one leaf (right) a transverse cut severs direct connection with the rest of the stem and hence axillary shoot grows and the leaf, which began to turn yellow, turns green again and its marginal buds are inhibited; the opposite leaf (left) turns yellow, its marginal buds develop and its axillary bud is inhibited.

ber of specific amino acids; the lower rosettes, which produce a great number of marginal buds, contained mainly carbohydrates.

The evident shift of chemical substances, together with the wilting of the lower leaves, can be ascribed to the inhibitory influence of the upper leaves, especially when the root system is not very active. It does not occur when the root system is fully active; then not only do the lower leaves remain alive but, if the upper nodes cannot produce vegetative shoots, new shoots develop in the axils of the lower leaves. This is especially true for all those graminaceous plants which develop many basal sterile shoots. These shoots can even weaken the main stem, making its development only partial and decreasing its productivity. Such a relationship could be called "basal dominance." This phenomenon is expressed in aging fruit trees by the growth of new shoots in the lower part of the crown while the upper branches die off. Thus the correlative conditions can change during the life of a plant, and apical dominance turn into basal dominance. The cause doubtless lies with the metabolic changes which occur within the plant.

Variations in the correlations which are responsible for polarity in plants can be studied in the following way. Pea seedlings are grown in the dark past the stage of the compound leaf, and then decapitated at the base of the fourth internode. If the cotyledons have been left on, the apex is always replaced by the growth of a shoot from the *uppermost* node, which bears a compound leaf. The result is the same in the light and in the dark (Fig. 33a and 33b). If both cotyledons are removed, and thus all reserve substances taken away, the uppermost bud (Fig. 33c) can develop only if the plants are exposed to strong light. In the dark, on the other hand, the buds which develop are the *lowest* buds, in the axils of the cotyledons (Fig. 33d); the upper buds will develop only if the cotyledonary buds are cut out. In the dark, therefore, there is basal dominance if there is a lack of reserve material, but light can restore apical dominance because it destroys the inhibitions exerted on the upper buds. This was seen above in the plants decapitated just above the cotyledons, one of which had been removed; in strong light the remaining cotyledon no longer inhibited its axillary bud. Indeed, this bud often grew more than the bud in the axil of the amputated cotyledon, although the latter always grows more in the dark (Fig. 16c).

As a corollary, the growth of any bud on a young pea or bean plant (*Phaseolus*) can be induced by exposing it to light while all the other buds are kept in the dark (Fig. 33g). The illuminated bud is thus liberated from the influence of the upper buds and acts itself as an inhibitor of those buds. It is for this reason that buds develop first on that side of branches or roots which receives the stronger light, and they inhibit the growth, or the regeneration, of buds on the other side. Trees branch out more on the side exposed to the stronger

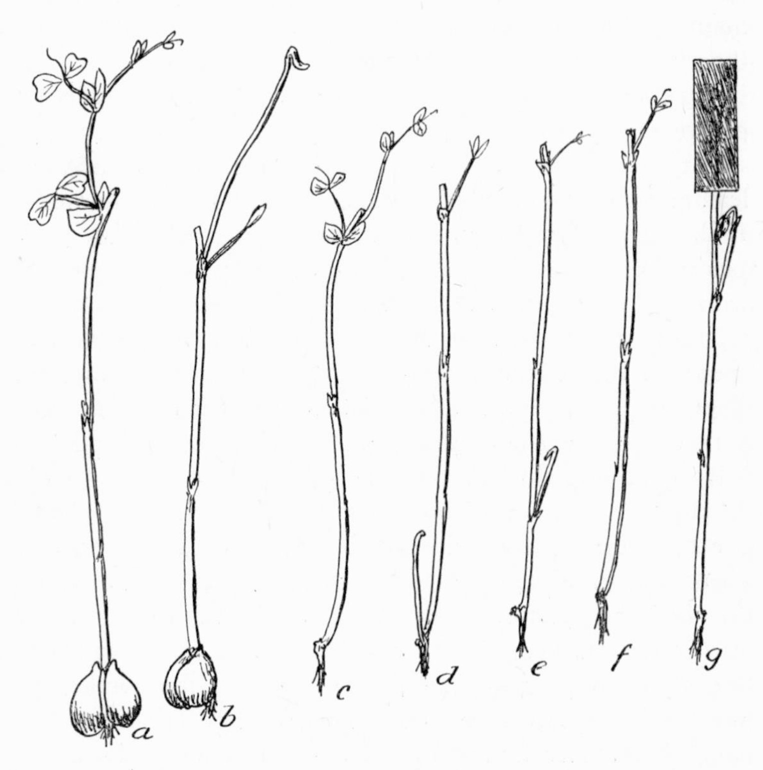

Fig. 33. Garden pea (*Pisum sativum*), stems grown in the dark and decapitated above the first true leaf: (*a*) in light, apex is replaced from the topmost node; (*b*) same result when plant is kept in the dark; (*c*) same result in very strong light when cotyledons have been removed; (*d*) in the dark, after cotyledons are removed, only cotyledonary buds develop; (*e*) when auxin paste is applied on the stub of the cotyledonary petioles, buds higher on the stem develop; (*f*) auxin paste applied to the topmost section inhibits growth of buds completely; (*g*) if topmost node is darkened, bud on next lower node develops.

light; branches can be induced on the shady side by cutting off those on the light side.

Fruit trees can be easily multiplied by exposing the upper part of their root system to light, when new buds and shoots develop there. Poplar roots close to the surface give rise to many shoots from random spots exposed to light, while such shoots never grow on roots kept in the dark. At the same time, by ringing the trees the movement of substances from the root to the base of the trunk can be prevented, and bud formation at the base promoted.

In the above-cited experiment with dark-grown pea seedlings, the cotyledonary buds on the stems of the plants without cotyledons regenerate the most readily (in the dark), because they are the oldest and most differentiated buds and yet are not inhibited by their cotyledons. For the cotyledons, with their reserve material, exercise an inhibiting influence, as has been proved in all the experiments discussed earlier; therefore they promote the growth of buds situated further away and thus bring about apical dominance.

In seedlings whose cotyledons contain few reserve substances, for example in those of flax, the cotyledonary buds grow out (even in the light) more than any other bud on the main stem, when the plants are decapitated high up, above the eighth or twelfth leaf. It is only when the stem is much further along in its growth that the dominance of cotyledonary buds over the other axillary buds eventually disappears. In those varieties of flax which bear more than one stem, this basal dominance is so pronounced that the cotyledonary buds develop even on intact plants, making these varieties branch out from the ground level.

If auxin paste is applied to the stubs of the cotyledonary petioles after the cotyledons have been removed, the buds in the axils of the cotyledons are inhibited while the buds in the axils of the scales at the next node develop, and these now take over the role of dominating the upper primordia (Fig.

33e). This happens probably because these buds were already prepared for such a possibility in the seed. On the other hand, if the same paste is applied to the stub of the decapitated main stem, the growth of all the buds is inhibited, so that the cotyledonary buds do not develop even if the cotyledons have been removed (Fig. 33f).

Basal dominance is a frequent phenomenon in plants. After all, the fact that growth is more active at the apical end is only one of the expressions of the polarity of the stem and it cannot be generalized. The correlations responsible for the polarity of organs are much more complex and are influenced by internal and external factors, such as nutrition and light.

We have already pointed out the practical importance of the fact that these correlative interactions can be changed by agricultural means, such as varying the distance between furrows, or their direction, or using different fertilizers. But, at the same time, the totality of the plant must be considered, because to influence a single organ and not the whole plant is meaningless. Thus, as far as nutrition is concerned, every part has to be considered with respect to its role in the plan as a whole.

Growth stimulators can be of great help in such an enterprise. If properly applied they permit profound interference with the correlative phenomena taking place in the plant. They can be used either to stimulate or to inhibit the growth and development of the plant, even though the biophysical and biochemical causes of morphogenesis have not yet been clarified. Morphogenesis entails a complex of reactions which have arisen, over the long period of phylogenetic development, from the interactions between external factors and the protoplasm. Many of the artificial stimulators or inhibitors are transported in the plant through the same channels and at roughly the same speed as the normal factors of auxinic nature.[4] Thus they also channel the transfer of other products

[4] Most of the rate measurements show that for movement from cell to cell the

of metabolism which move from the leaves to the sites where they are used up. Many correlations are involved in this process, as we have seen in experiments reported earlier in this book.

Quite remarkable also are the results of experiments by Chailakhian (1956), who followed the migration of the products of assimilation from leaves which had been given radioactive $C^{14}O_2$ for twenty minutes. He showed that the products first spread from the leaf to the bud in the axil of that leaf, then, and to a lesser degree, into the stem and roots and, last, and least of all, into the opposite axillary bud. However, experiments described earlier show that the growth activity in the plant is not directly linked to this migration of the products of assimilation. The growth activity is governed by hormone relations and therefore the bud in the axil of the opposite, amputated, leaf is the one that grows the most. This is true just as much in the case of cotyledons as in that of fully grown, intensely photosynthesizing leaves. That what is involved is not merely the influence of the products of assimilation but rather specific regulation is shown even better by the fact that seedlings of flax behave in exactly the opposite way, that is, the bud in the axil of the remaining cotyledon is the one that grows the more.

This regulation becomes particularly clear in experiments with pea or bean seedlings which have been decapitated just below the first primary scale. Indoleacetic acid, triiodobenzoic acid, gibberellin, and maleic hydrazide were applied either on the cut surface or on the intact epidermis in the middle of the internode (Fig. 34). The cotyledonary buds grew out in all cases except for the plant treated with auxin; in this plant a large callus with roots developed under the layer of auxin paste, and the buds were inhibited (Fig. 34a). In the other

---

synthetic auxins move only at 5 to 50 percent of the rate of indoleacetic acid. Movement in the xylem is, of course, independent of the auxin used, since it depends primarily on the rate of transpiration. (Ed.)

cases, the shoots had to be removed from time to time to prevent the exhaustion of the cotyledons. The experiment was terminated when, on the plants treated with auxin paste in the middle of the internode, the part above the treated zone withered (Fig. 34b). At the time, the stems which had auxin

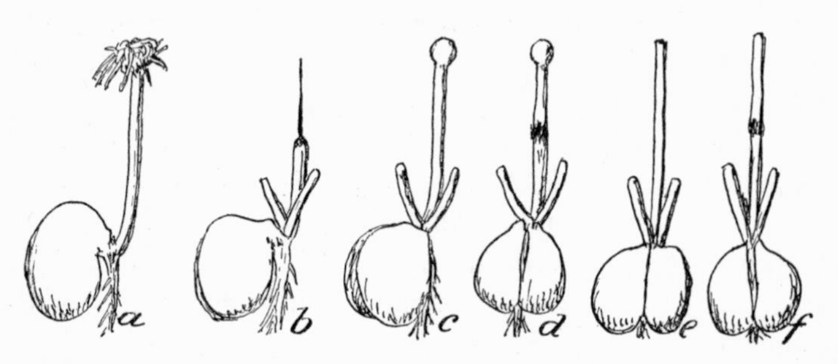

Fig. 34. Garden pea (*Pisum sativum*), young seedling decapitated below the first primary scale: (*a*) auxin paste applied to the top section produces a tumor with roots and inhibits cotyledonary buds; (*b*) auxin paste applied to the middle of the stump causes the stem above to die off; (*c*) triiodobenzoic acid paste applied on the topmost section, or, (*d*) in the middle of the stump, causes the same swelling of the apical end, but without roots; (*e* and *f*) maleic hydrazide paste applied to both places produces neither swelling nor withering. The growth of axillary buds is not inhibited in *b* to *f*.

on the top section were still completely healthy. All the plants treated with triiodobenzoic acid behaved in the same way regardless of the site of application (Fig. 34c and d). Their apical ends produced spherical swellings but no roots; the cells merely multiplied and enlarged. This response may be ascribed to translocated organic materials, carried upwards by the triiodobenzoic acid, which spreads in all directions but mainly from the base toward the apex. Similarly the withering of the part of the stem above the ring of auxin shows that auxin, which travels in the plant from the apex toward the base, "carries with it" also other nutritive substances. Thus the upper portion becomes starved and dies.

Gibberellic acid and gibberellins in general behave in the

same way as triiodobenzoic acid in this experiment, except that the swelling produced at the upper end is quite small or even unnoticeable. These substances can also spread upwards but they are probably incapable of bringing about the growth of cells in the transverse plane, since the cells only become greatly elongated. On the other hand, indoleacetic and triiodobenzoic acids show, in this simple experiment, a marked difference in the direction of their transport in the plant, a difference which is in accord with the well-known difference between their respective influences on the determination of form.

Because auxin strongly promotes the formation of roots, it is used in practice to hasten the vegetative reproduction of cuttings. From experiments with tuber-forming plants, auxin also promotes to some extent the formation of reserve organs. Triiodobenzoic acid, on the other hand, does not stimulate the formation of roots but in proper concentrations causes the elongation of stems, and in many families such as the Leguminosae, Liliaceae, and Cruciferae, it increases and speeds up the production of flowers and fruits. This action may be due to its ability to travel upward in the plant and thus to promote the translocation of plastic substances to the top, where they can be used in the formation of flowers and fruits. This was first shown in our country [Czechoslovakia] by Vaníček (1950) on the tomato and cauliflower and by Šebánek (1952) on the sunflower.

Plants which in the same experiment were treated with maleic hydrazide (2 percent) showed no change during the course of the experiment (Fig. 34e and f); the stub neither swelled nor died, the section remained smooth, and the buds in the axils of the cotyledons grew out. Evidently MH does not bring about a detectable migration of substances, nor does it act as an inhibitor under these conditions.

The differences between the effects of these most common synthetic stimulators correspond to definite mechanisms which assure the integration of plants during their development. But it would be dangerous to represent such mechanisms in too simple a fashion, because the basic regulatory factor is not actually any one of these substances or their analogues in the plant, that is, native auxins or inhibitors, but is rather the living matter itself, on which these substances act. A simpler object, such as the acellular alga *Caulerpa prolifera*, was therefore found more appropriate for the study of cellular dynamics.

If we isolate a leafy assimilator from a healthy *Caulerpa* plant, the base soon loses its chloroplasts, becomes colorless, and is transformed into a sort of physiological substitute for the lost rhizoids. These actually are regenerated, but only on the second or third day. If both poles of the assimilator are subject to the same external conditions, the rhizoids always regenerate at the basal pole. However, if the assimilator is divided transversely into two halves, one apical and one basal, rhizoids will grow at the basal end of both of them, thus complementing the green assimilator in order to form a new unit capable of existence and further development. Later, these cuttings develop "leaves" or stems, whose localization and nature is controlled by the external environment.

If the basal half (which has the lowest level of inhibitions) is planted in sand or left in water it forms only a few rhizoids, and the new assimilators or rootstocks appear very close to the base (Fig. 6e). If, on the other hand, the apical part, the top of the plant (which is the source of inhibitions), be planted in soil enriched with the mud of the original culture, a large number of highly branched rhizoids appear, and new assimilators regenerate immediately below the tip (Fig. 6d). The great development of rhizoids acts here in the same manner as triiodobenzoic acid acts on the pea plant, namely to push the zone of growth upward. It is therefore probably in the rhizoid that the substance responsible for apical dominance arises, a situation that also holds in the roots of higher plants. If there are only a few rhizoids, basal dominance is expressed, probably because of the inhibitory influence of the upper segment of the assimilator. We have here a faithful analogy with perennial higher plants which, at the end of the vegetative period, restrict their growth to the underground organs because of the inhibitions operating in the aerial parts.

In higher plants the upward flow of plastic materials, under the influence of triiodobenzoic acid, brings about the formation of flowers and fruits; on the other hand, the downward flow of these substances under the influence of auxin stimulates the development of vegetative organs of multiplication. Thus there are apparently two poles of differentiation, an apical pole of floral differentiation and a basal pole of tuber differentiation, as in *Scrophularia nodosa*. One of the factors contributing to this is the action of the leaves in inhibiting the growth of their own axillary buds, so that their assimilation products can be used only in places which are as far as possible from them: the poles.

This inhibitory effect of the leafy stem can be demonstrated with a pair of branches growing on the same node. These are isolated from the roots, together with a part of the axis but without their supporting leaves, and from one branch all the leaves are removed. It makes no difference whether the smaller or the larger branch is the one defoliated; in either case, subsequently the branch with the leaves stops growing, while the branch without them continues to develop (Fig. 15d). The nature of the growth depends on the site of the maternal plant from which the branch was taken; a section taken from the base forms tubers, while one from near the top forms flowers. The tip of the leafy branch stops growing and the plastic materials migrate across the node to the leafless shoot. The action does not depend on the height of the node. On the leafless shoot tubers can form on higher nodes and flowers on lower nodes, by comparison with the shoot which has leaves.

A fully leafless stem behaves in exactly the same way as a branch bearing leaves. Its poles correspond to the leafless branch, and its flowers are formed sooner, while the stem is still rapidly growing. On the other hand tubers appear only much later, after the stem has stopped growing. If the leafy stem of a Scrophularia plant, at the end of its flowering period, is cut transversely into two halves of equal length, the upper half forms floral shoots at the top and the lower half develops tubers at the bottom as if the stem had never been divided. On the other hand, bud development is almost completely inhibited in the axils of the largest leaves near the cut. The flower primordia on the upper half of the stem are probably centers of attraction for all the plastic materials produced by that half, just as the tubers are centers of attraction for all the translocated substances of the lower half, preventing the growth of buds on the newly created upper pole.

This type of polarity is very different from that which governs the formation of roots, for it results from inhibitions caused by the leaves and therefore depends on the nature of their products. It is quite easy to cause the formation of tubers at the top of a stem instead of at the base. For this purpose the apical primordia are immersed in water, and the base is planted with the maternal tuber also in water. Both ends are then inserted into corks in wide test tubes. Thus we have the same conditions at both ends and therefore we get the same formations. The young apical bud primordia stop growing, the axis enlarges a great deal, and from the nodes many roots develop. This is the same thing that happens when tubers are formed (Fig. 11a). The results of this experiment are important, because in this case all the leaves, even the topmost one which usually produces substances necessary for the formation of flowers, collaborate to form tubers. If we take isolated pairs of leaves from the upper part of the stem, only buds are formed if they are kept in the air (with their bases in water); if immersed in water, however, the buds also change into tubers. Therefore differentiation is governed not only by the quality of the leaf products but also by the external conditions under which the experimental organ is kept.

Conversely, flowers can be made to form at the base of the stem or even on the maternal tuber. This has been done many times with other plants, by removing all the flower primordia on the stem (for example, by Mattirollo, 1900, with pea plants). In Scrophularia it is necessary to cut out all the axillary buds and the apical bud. The experiment is successful only during the period between the end of differentiation and the beginning of the unfolding of the bud (Fig. 11c). A similar operation performed later, when the bud has already begun to unfold, will only accelerate the development of daughter tubers. The first prerequisite for the development of flowers on the mother tuber is the presence of the so-called floral stimulus (which means that the plant is in other respects in the flowering state). Because of the activity of the flower primordia all plastic materials are being drawn toward the top. The young flowers are centers of the formation of auxin, which causes the elongation of the upper internodes and inhibits the growth of leaves. If the stem is cut off above the lowermost primordia a shoot grows from the bud with exactly the same number of leaves as on the first stem. Therefore the complete transformation of the lower buds, even those on the mother tuber, into flowers is brought about by the presence of the floral substances in the whole stem, from which all of the axillary primordia have been removed.

By contrast to what occurs in Scrophularia, most trees accumulate reserve substances in the parenchyma of their roots and branches, and the flower primordia (as, for example, in the lilac) are formed in the upper part of the plant. In some rare cases inflorescences occur at the base of the shoot, showing that even the base can be a suitable place for the formation of flowers and buds. Such formation is determined during embryonic development, so that above the scales floral shoots are formed first and only after that the green assimilatory leaves. What we bring about in Scrophularia by cutting out all the buds on the stem occurs in the lilac spontaneously (under conditions still unclear) during the embryonic development of the previous year.

In any case, whether we are concerned with the polarity which determines the formation of roots, or the growth and regeneration of buds, or the formation of tubers and flowers, the influence of the axis is always decisive. The best example of the polarity of roots is provided by the strong roots of dandelion (*Taraxacum officinale*) or of chicory (*Cichorium intybus*) which regenerate buds at the basal pole. If a thick section is cut out from one side of the root, buds will regenerate not only on the basal section of the root, but also from the scar under the removed sector, while the apical part of the root remains unchanged (Fig. 31d). The buds regenerate from primordia formed on the callus. A perhaps comparable regeneration of the tentacles of the sea anemone, *Cerianthus membranaceus*, occurred in an experiment performed by Loeb (1906). Here he cut out a sector from one half of the body and noticed that at the oral pole fewer tentacles developed on the side above the section than on the other side (Fig. 31e), while on the scar left by the lower edge of the sector new large tentacles developed. It is also interesting to compare the regeneration of buds of horse chestnut (Aesculus) described by Němec (Fig. 31f), with that of Planaria, studied by Morgan (Fig. 31g). In *Cichorium* a great number of buds develop, even on longitudinal sections.

Vöchting (1892) discovered that tuberal roots show a so-called radial polarity. For example, he cut out a cube from a sugar beet and replaced it in the wound. If he kept the original orientation the cube joined again with the rest of the root tissues, but if he reversed the orientation so that the external tissues of the cube were on the innermost surface, a large callus formed between the adjacent surfaces and pushed the cube out. Radial polarity is apparent also from the fact that a thick root grows progressively in width by the addition of new layers through the successive activity of the cambium. It is also evidence for the radial migration of specific substances.

Given these facts we can expect in the same cases the existence of a tangential polarity (Fig. 35). If a longitudinal

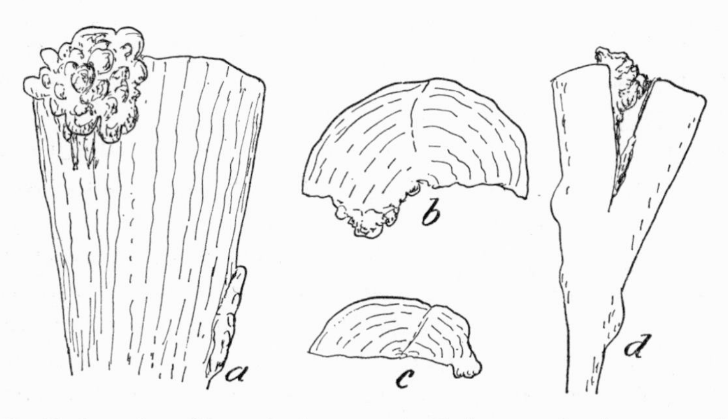

Fig. 35. Sugar beet (*Beta vulgaris, ssp. saccharifera*), a root showing tangential polarity: (*a*) on longitudinal section, tumor grows at top left near the center and at bottom right on the edge; (*b* and *c*) transverse sections through these tumors; (*d*) cut into decapitated root with tumor.

section is made through the spherical part of the beet root and the changes which take place on the cut surfaces are observed, we first see that a large callus is formed near the central pith on one side near the top, while the other side remains smooth at this level (right side of Fig. 35a, section shown in Fig. 35b). A callus may also form on the outside, over the sectioned outer rings. This happens only at the right lower side of the section (Fig. 35c). In both cases, therefore, the callus is formed in a place where the vessels which ran obliquely through the spherical part of the root have been cut at their bases by the longitudinal section. Due to the divergent orientation, this occurs on the top left side in the center of the section and at the lower right side on the outside edge of the section. In reality, therefore, this would be a case of longitudinal polarity but for the fact that the vascular bundles are deflected (by the contracted growth of the roots) from their vertical course, the deflection varying from layer to layer. Both sides are always equally subject to gravity, which is therefore not responsible for the difference in the development of the callus.

The root tubers of *Corydalis solida* also regenerate roots at the bottom and buds at the top. Similarly the potato tuber, if all its buds are cut out, regenerates buds with the same polarity as the stem, that is to say, at the morphological top, roots being formed at the base (which is here the end nearest the parent stem).

When the roots are removed from a tuber of *Scrophularia nodosa* new roots regenerate all over the surface of the tuber, but especially at the apical end. If the last daughter tuber is separated from the mother tuber, then the former forms roots also on its basal section (Fig. 23a), but in the middle where the tuber is widest in diameter roots do not appear, perhaps because they are inhibited by the high content of reserve substances. A similar polarization is found in runners which also show the greatest development of roots at their apical ends. Buds arise near their upper pole, although branching occurs as in the roots further away from the top. On the other hand the leaves stop growing first at the apical pole and only some time later at the base.

A comparable movement of specific substances can be demonstrated in the needles of the yew tree (*Taxus baccata*). If agar containing auxin is applied to the apical cut surface, then pure agar applied to the basal cut surface soon shows the presence of auxin in relatively high concentration. In the reverse experiment auxin is not found in the agar at the apical surface. Therefore auxin travels (perhaps through vascular bundles) from the apex to the base, and in the places where its movement is blocked roots arise. Also, stems of limited growth like those of *Tussilago farfara*, if isolated, form buds in the axils of lower scales. On the other hand, leaves which grow for a long time at the top of a plant, as in *Utricularia longifolia*, form adventitious buds at the top margin because meristematic tissues remain there.

Finally there are leaves, like those of the Bryophyllum group, which contain traces of meristematic tissues widely distributed over the surface. These are interesting because of their deflected polarity. Since the tip of the leaf is the oldest part, the marginal buds there are the most differentiated and these inhibit the development of the buds situated on marginal dents lower down. These marginal buds are very helpful in the study of the correlations which exist between leaves, for such correlatives are otherwise hard to elucidate. However, they can be very well investigated in the leafy assimilator of the acellular alga *Caulerpa prolifera*, which (as noted above) shows polarity similar to that of multicellular higher plants. Isolated assimilators form rhizoids first of all, and later either leaves or a rootstock, at the base. If supported upside down, their polarity is not altered; all the organs still develop at the base, which is now pointing upwards. This strong polarity is due to the inhibitory influence of the apical edge, which makes itself particularly felt in the light.

The assimilator can be divided into an inner and an outer part, both of which will then regenerate independently even when they remain connected through a thin layer containing mainly vacuole. In the normal position both parts regenerate in the same way at their base (Fig. 6b); but in the inverse position there is a great difference in their polarity. The outer part maintains its original polarity and regenerates at the base, while the inner part reverses its polarity, probably under the influence of gravity, to form both rhizoids and leaves on the lower edge, in reality the apical edge of this section (Fig. 6c). Gravity therefore cancels the polarity of the inner part in which the streaming of the protoplasm is much more intense; at the same time this part is freed from the inhibitory influence of its edge. The decisive influence on the form of the plant is the prerogative of the outer layer of the protoplasm, but the rest of the protoplasm which flows all through the alga, and is probably most affected by gravity, also participates. It is sometimes erroneously assumed, following F. Steinecke, that when *Caulerpa* assimilators are placed inversely the apical protoplasm passes upward and the basal protoplasm downward, the inner polarity remaining unchanged. As is true for higher plants, new primordia are formed in all cases, corresponding both to the growth correlations and the external conditions.

Another equally important influence on the polarity of the alga *Caulerpa prolifera* is light. (In this respect Caulerpa behaves in the same way as the other frequently studied member of the Siphonales, Bryopsis.) If a leaf or a rootstock is illuminated on one side only, then this side will form leaves or rootstock, while rhizoids will appear on the other side. The leaf surface shows a dorsal and a ventral side. Blue light is most effective and red light quite ineffective, acting simply like darkness. If an isolated leaf is illuminated through a blue filter on one side and a red filter on the other side, new organs are formed only on the blue side, even though the total amount of light transmitted is less.

Perhaps blue light stimulates a greater production of proteins, which activate the growth process, while in red light mainly carbohydrates are produced and these inhibit the regeneration of assimilators. Light has therefore a similar effect on *Caulerpa prolifera* as on fertilized zygotes of algae where, as we have seen, polarity is determined primarily by the influence of light and gravity.

Both of these basic factors, which influence the growth and movement of plants, promote integration in plants. Gravity enables roots and rhizoids to penetrate into the soil, light enables the growth of assimilatory organs upwards toward the source of light. The same is true for plant movement in response to these factors, namely geotropism and phototropism. The interaction of these two antagonistic influences results in the formation of the plant as an integrated whole.

Polarity is interesting also from the biochemical and biophysical point of view. There is a gradient of substances in the plant, for, as Molotkovsky (1954) found, a difference in the water content is detectable not only between the base and the apex of the whole plant but between the two ends of each internode. A similar gradient occurs in enzyme activity (for example, catalase) and in chlorophyll content. This phenomenon is important for understanding the typical structure of plants, which comprise consecutive metamers. Although the metamerization is determined in the embryo, it is expressed only in the adult state in the variety of leaf shapes and anatomical structures, as formulated in Zalensky's law (1904).

CHAPTER VI

Transplantation and Integration

Transplantation such as grafting, which is an ancient method of vegetative propagation, shows how isolated sections of plants can once more become integrated in a functioning whole, not by regenerating missing organs themselves but by utilizing those of another plant. The scion and the stock grow together and correlations are established between all the organs of the new whole. If the scion and the stock differ in their metabolism, one of them can alter the other. Their properties can also change, as was excellently shown by Michurin (1948), especially in the case of the so-called vegetative "rapprochement" (reunion) and the influence of the "mentor."[1] The alteration of physiological properties can be revealed by changes in color, shape, taste, precocity, durability of the fruit, resistance to frost, and other characteristics of fruit trees. By grafting one can also cross plants which cannot be crossed otherwise, like the pear and the mountain ash.

Especially interesting are transplantations which bring about biochemical changes, for these may sometimes direct our attention to the site of formation of certain substances.

[1] These views are in general no longer held outside the USSR. (Ed.)

Reciprocal grafts between alkaloid-producing plants and plants free from alkaloids have been particularly enlightening in this respect, and have led to the discovery that alkaloids originate mainly in the roots. If *Nicotiana tabacum*, which contains nicotine, is grafted on *Nicotiana glauca*, which produces anabasine instead of nicotine, the leaves of the scion contain anabasine characteristic of the stock, rather than nicotine (A. A. Chtuk, D. Kostov, and A. Borodinova, 1939). Later Chtuk experimented with grafts of tobacco and tomato. If the tobacco was grafted on the tomato, its nicotine content became practically undetectable; if, on the contrary, the tomato was grafted on the tobacco, an appreciable amount of nicotine was present, even if all the leaves of the tobacco stock had been removed. It follows that nicotine is produced in the roots and rises up to the leaves, regardless of whether these belong to a tomato scion or to a tobacco plant.

Roots are therefore not merely absorptive and anchoring organs, but they represent special chemical laboratories, producing in particular organic nitrogenous substances (Mothes, 1956), which stimulate and rejuvenate the plant. Before the discovery of the true plant growth factors, some investigators (for instance, Cavara, 1923) saw in these root products, especially in the alkaloids, analogues of animal hormones. Alkaloids are plant products, and animals, which do not synthesize them, use them as medicine.

Actually alkaloids do not always originate in roots — solanine is formed in the leaves of the potato, anabasine in the leaves of *Nicotiana glauca*, and quinine in the leaves of Cinchona — but in general roots are more active in this regard than leaves. Atropa and Datura, for example, form atropine in their roots, and it can appear in other plants grafted on these roots. The sweet variety of Lupinus or of Pisum, when grafted on the bitter variety, contains bitter alkaloids in both leaves and fruits. This phenomenon is seen also in rubber plants; if a scion of *Taraxacum officinale*, which is free from rubber, is grafted on to the roots of *Taraxacum kok-saghyz* (the rubber-forming Russian dandelion), the scion will contain rubber. Similarly, Rudbeckia does not produce rubber, but when grafted on *T. kok-saghyz* it does contain it.

Certain substances spread from the roots of the stock to the leaves of the scion and are there transformed into alkaloids. This transformation is not a specific process, for it can be carried out by plants which do not normally produce alkaloids. It has been stressed above that roots have the capacity to transform mineral nitrogen into amino acids, as was shown by Kursanov (1952); they can do this even if they themselves do not take up nitrogen from the soil. If a nitrogen-containing fertilizer is applied on the leaves, the nitrogen migrates into the roots where it can be used up. The same is true for the formation of alkaloids in roots. Roots which for weeks have not taken up any nitrogen from the environment still contain alkaloids in the guttation liquid. Scions of Atropa on a tomato stock, spread with a urine solution as an appropriate source of nitrogen, did not contain any atropine.

These phenomena are very important in the life of the root system and therefore of the whole plant. It is easy to see how crucial is the choice of the stock, especially with fruit trees, for it can alter the intensity of growth, the longevity, the fertility, and other characteristics. Michurin showed how a wild stock can degrade the quality of a cultured variety and, vice versa, how a wild variety can be bettered by a cultured stock.[2]

The experiments of F. W. Went (1938) with embryonic pea plants may be thought of as a morphological model of these phenomena. Went tried to show by transplantation that in the roots there originates a specific substance indispensable for stem growth, which he called Caulocaline. By contrast to auxin, this substance was thought not to be able to penetrate

[2] See note to beginning of this chapter.

through a non-living medium, such as a layer of gelatin or agar, but to require that the grafted parts grow together before it could be transported across the graft. It is generally felt, however, that correlative interactions and the transformations of substances they involve cannot be treated in such a simplified manner. This is especially true for the determination of specific shapes, such as the formation of flowers and tubers. The power to form these organs can also be transmitted by grafting.

Razumov (1931) experimented in this field with tubers of the wild species of the potato: *Solanum leptostigma*, *S. semidemissum*, and *S. antipoviczi;* these varieties come from equatorial regions of South America, where the days are short, and therefore do not form tubers under the long days which prevail in his country[3] in the summer. When, however, leaves of our common potato *S. tuberosum* were grafted on them, they formed tubers even under summer conditions of long day. However, scions of other Solanaceae had the same effect even if they came from plants which do not produce tubers, such as *S. nigrum*, tomato, Atropa, or Datura. Only the leaves of the tobacco and of Physalis failed to bring about the formation of tubers when grafted on the wild varieties of the South American potato.

This observation probably cannot be generalized, but nevertheless the assimilatory products of leaves of a large number of plants can in many cases transform bud primordia at the base of the plant into tubers, which would not have appeared if grafting had not been performed. The determining influence resides therefore in the assimilation products of the leaves, usually those from other plants, which are not being used up in the scion, but migrate into the stock and there, in spite of adverse conditions, induce the formation of tubers. Since tuber production was obtained with such plants

[3] Leningrad area.

as *Solanum nigrum*, tomato, or Datura, it follows that the assimilation products do not contain any specific tuber-forming substances. This was shown previously in Lindemuth's (1901) experiments: A potato scion grafted on a Datura stock grew very well and formed, above the stock, horizontal runners which suddenly, under the influence of vigorous root development in the stock, turned into leafy stems. When, however, the red pepper was used as stock, the scion grew little and instead of runners gave rise directly to tubers at the same point.

The formation not only of tubers but also of flowers is transferable by grafting, that is, from a flowering stock to a non-flowering scion. Vöchting (1892) thus grafted a vegetative bud of the sugar beet on a two-year-old beet root, and observed that it changed into an inflorescence; on the other hand, an inflorescence grafted on a one-year-old root changed into a vegetative stem with large leaves. One-year-old specimens of biennial plants such as the sugar beet or the carrot, grafted on to second-year plants capable of forming flowers (or on to plants which flower in their first year), can themselves develop flowers. Late flowering varieties grafted in the spring show comparable behavior (cf. the changes shown in Fig. 22a and b). One has to remove all the bud primordia from the stock and all the leaves from the scion, however, since otherwise plants grown under long-day light conditions would remain crouched in the form of leafy rosettes near the ground.

Both the short-day variety of Perilla (*Perilla nankinensis*) and the common soybean form flowers even under long-day conditions when grafted on to flowering plants. Other examples are: *Helianthus tuberosus* (Jerusalem artichoke) grafted on to the sunflower; the short-day Maryland mammoth variety of tobacco grafted on to a day-neutral tobacco of the Samson variety; a one-year-old biennial Hyoscyamus grafted on the root of an annual Hyoscyamus; and a long-day variety of

Bryophyllum grafted on to a plant which had been made to flower under short-day conditions (Fig. 21b).

The experiments of Chailakhian and of Moshkov, in 1939 and later, show clearly that one can transfer the flowering stimulus or "florigen" to a non-flowering plant by grafting

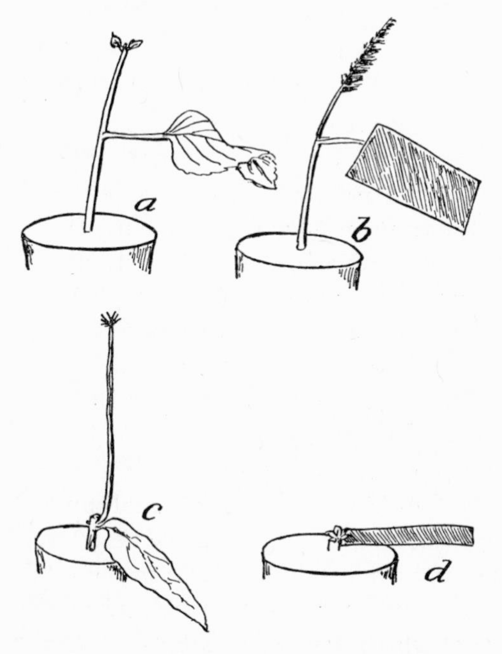

Fig. 36. *Perilla nankinensis* (= *Perilla frutescens*, var. *crispa*) (short-day plant): (*a*) leaf in long day, tip vegetative; (*b*) leaf in short day (covered with black paper for part of each day), tip forms an inflorescence. Spinach (*Spinacia oleracea*) (long-day plant): (*c*) leaf in long day, flowering stem develops; (*d*) leaf in short day, tip vegetative (after Chailakhian).

just one leaf exposed to the right photoperiod (short-day for the Perilla and long-day for the spinach, *Spinacea oleracea*) on to the non-flowering plant (Fig. 36). The crucial role of leaves also becomes evident from experiments such as those in which an isolated leaf without buds, exposed to the appropriate photo period and grafted onto a non-flowering stock, causes the stock to flower. One leaf therefore is enough[4] to change the development of a plant from the vegetative to the reproductive stage. This involves a deep physiological transformation; the activity of the cambium ceases and the one-year-old plant dies after having produced flowers.

There is no reason to doubt[5] that during transplantation the transformation of both partners is so deep that not only do they themselves show new characteristics in their individual lives, but even their seeds may be affected and transmit new characteristics to the progeny. According to Michurinian genetics, heredity is not linked to only a few specific elements of the cell, but is carried by all of its living components, and further it can be influenced by the external environment. Such a great deal of evidence has already been assembled to show the profound changes in metabolism caused by grafting that we can consider grafting as a means of vegetative hybridization. The work of the French grower Daniel (1921) especially supports this idea. A recent critical review by Glushchenko (1957) brings out the value of all the experiments on vegetative hybridization in plants and animals.

Darwin proclaimed that the basis of heredity lies in correlations. The problem resides not only in the establishment of correlative interactions between stock and scion, but also in the manner in which these correlations are transmitted, that is, via sexual cells or via somatic cells, both of which give rise to a new plant. A very important task of molecular biology is to solve the question of what concrete changes occur not only in the DNA and proteins of the chromosomes, but also in those other cell components which participate in expressing the hereditary characteristics. We shall thus slowly arrive at an understanding of the physiological mechanism of sexual

[4] One eighth of a leaf sufficed in the experiments of Hamner (1942). (Ed.)
[5] Most Western workers think there is grave reason. (Ed.)

hybridization and of heredity in general, in the broadest sense of Timiriazev (1892).

Darwin himself hoped that vegetative hybridization would cast light on sexual hybridization. This prophetic view was at that time founded on the occurrence of the so-called "chimaeras," such as *Cytisus adami*, arising from the union, by growing together, of *Chamaecytisus* (*Cytisus*) *purpureus* and *Laburnum* (*vulgare*) *anagyroides*, or such as *Crataegomespili*, due to the similar union of *Cratageus* and *Mespilus*. Microscopic preparations, especially those taken from the flowers and fruits, show that in these unions the individual tissues of the two plants remain separate, but as one plant grows around the other there is a marked change in the shape of leaves, the shape and color of fruits, and in other organs.

Winkler (1907) obtained experimental chimaeras between *Solanum nigrum* and the tomato. The tissues of the two plants grew together so closely that a single apex arose and gave rise to a composite plant. The two partners can either remain simply adjacent and the organs keep their original shape — these are sectorial chimaeras — or one partner can grow all around the other, producing the very interesting periclinal chimaeras. It is remarkable that two completely different plants can thus fuse into one harmoniously functioning whole. The two partners of a periclinal chimaera only give themselves away by forming adventitious buds, which comprise tissue of just one of the partners. There are cases in which adventitious buds arising exogenously from the external layers of the stem, and those arising endogenously from the internal layers of the roots (the pericambium), produce completely different plants.

The organic union of a periclinal chimaera is due to the correlations established between its tissues and cells, which belong to different plants. One part, for example, can produce the epidermis and no inner tissues, while the other part produces the so-called heart of the chimaera, that is, the pith and vascular bundles, and no epidermal or bark tissues. These components are thus correlatively inhibited. Close study has shown that the two partners influence each other to such an extent that we cannot really speak of a separation of individual layers.

CHAPTER VII

Integration and the Problem of Periodicity in Plants

The correlations between its organs which give the plant its organic unity (integration) are also responsible for the periodicity in the plants' responses. They determine when in the life of the plant a given structure appears, as well as when growth will stop and start again after the rest period. The rest period itself, in perennial plants of moderate or cold latitudes, appears to be a consequence of correlations which result from adaptation to the alternation of seasons and to low temperatures in the winter. By the late summer, trees in temperate climates have stopped elongating, owing to a steady increase in specific inhibitions which originate in the leaves, as we have seen in seedlings. Such a correlative influence is clearly demonstrated by some of our trees, which can bud out spontaneously several times in one vegetative period, to form the so-called summer shoots,[1] which are most obvious on the oak.

The rhythmical budding is a consequence of complicated correlations in the plant as a whole. The main role is played by inhibitions exerted by the leaves, which begin as soon as the leaves start to unfold in the spring; they inhibit the growth

[1] Lammas shoots, or "proleptic" shoots. (Ed.)

of the leaf primordia in the buds, which thus become smaller and smaller as we reach the tip of the plant. The topmost ones are as tiny as scales; the difference between them is shown by their withered blades borne between a pair of needlelike stipules. Only when the leaves are completely unfolded do their own scales appear right above them, as a pair of rounded stipules without a blade. Above these scales new foliage leaves can form. When in the terminal bud all the scales and leaves have been formed, the bud can grow out during the same vegetative period, forming a new branch; this happens regularly in the oak at the end of June. Sometimes the summer shoot formation can be repeated again, or even twice, if the fall is suitably mild and humid (Ille, 1937).

Klebs (1914) explained periodic budding in the oak and beech (Fagus) by assuming that each outburst of budding exhausts the nutritive mineral content of the soil in the vicinity of the roots, and that a certain lapse of time is necessary for the minerals to accumulate again or for the roots to penetrate to them. The root activity undoubtedly determines both *whether* summer shoots will be formed and *how many times*. In experimental cultures of potted plants, we have observed that when the root system was highly developed, through treatment with fertilizers, the plants formed summer shoots two or three times, while plants with weak root development produced only one shoot or none at all. On young trees summer shoots appear on all branches, while on older trees they appear only on the strong terminal ones and not on the weaker laterals. At the beginning of their growth these shoots are noticeable for the yellowish or even reddish color of their leaves.

Terminal shoots communicate directly with the roots, unlike lateral branches, which are dominated by the terminals. The roots stimulate the apical buds or terminal shoots, which then inhibit the lateral buds and shoots.

The question of whether the root system is really responsible for the formation of summer shoots can be finally answered only by experiment. For example, if we transfer the plants to darkness at the beginning of shoot formation, they will show continuous growth for a long time. Their leaves will remain white and small and have practically no inhibitory influence. Although they unfold completely they can inhibit only their own axillary buds; since they have no influence on the apical leaf primordia and cannot induce the formation of scales, the development of foliage leaves is continuous. The same results could be obtained in the light by progressively removing all the leaves which have unfolded.

Periodic growth therefore occurs only when the leaves develop freely in the light. When the size of the blades is around 8 to 10 mm, the leaves are already able to inhibit the apical meristem, although this is long before they are fully grown and photosynthetically active. Thus if a small segment of a branch, consisting of one node and one leaf with its axillary bud, is set in water with its single internode pointing downward (Fig. 37), roots will grow and branch

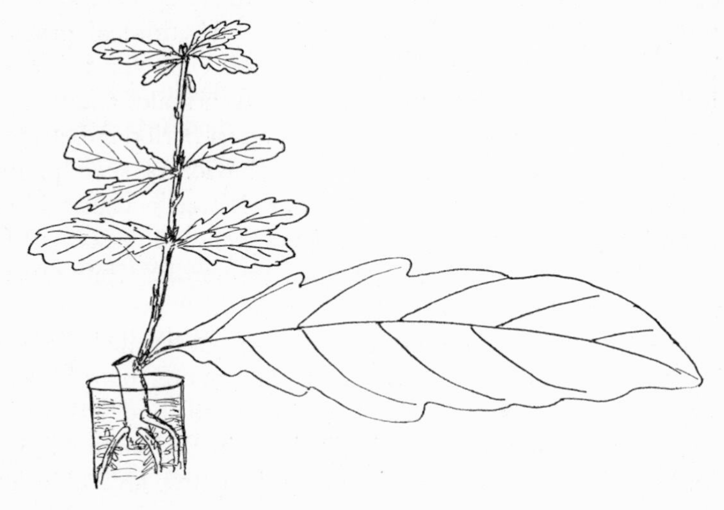

Fig. 37. Winter oak (*Quercus petraea*): segment kept in water shows three successive summer shoots, formed successively on III/23, VI/6, and VII/27.

out from the section, while the bud unfolds and forms leaves which inhibit the apical growth of the shoot. Then a new bud arises with scales and primordia of new foliage leaves. The embryonic formation of this bud takes time. In nature it requires 6 weeks to 2 months; in experimental cultures with continuous strong artificial light 2 weeks is sufficient. In reality therefore the segment is not resting, but is laying the foundation of the summer shoot which starts to unfold as soon as it is completely formed. In an experiment started in January, the first shoot was fully formed on III/23, the second on VI/6, the third on VII/27. Following transfer outdoors, during the summer months, two more shoots were formed.

Since the roots were only in water, the formation of shoots could not have been determined by the nutritional supply from the roots, for in these conditions this supply is constantly decreasing. Furthermore the humidity and temperature did not vary enough in the room to account for the rhythmical growth, and indeed the rhythm was greatly stimulated by continuous light. The effect of light cannot be explained in terms of a long photoperiod, for similar rhythmical growth is known for some trees in the tropics, even for the oak. Indeed, because the oak in the tropics forms scales and foliage leaves periodically, it appears to enter a rest period in between times. Our cultures resembled the behavior of such plants in the tropics. The plants in the pot showed certain differences in the rhythms of individual branches; some seemed to be resting while organizing new bud primordia, while others were actively burgeoning.

This localized variation in growth does not parallel changes in the activity of the root system. The action of the roots appears to be rather to stimulate rhythmical bud growth in a general way by lowering inhibitions, in the same way as does continuous light. Continuous light is less favorable for photosynthesis than intermittent light. The frail consistency of leaves and their pale green color is also an indication of weak inhibitions, by contrast to the dark green, sturdy leaves, grown under natural conditions in the spring. This may also explain why even on large oak trees most of the buds do not grow into summer shoots, but rest until the following spring.

Other woody plants, such as the maple (Acer), produce similar "proleptic" shoots, which after a brief decrease of growth rate of the apex, coupled with the appearance of short internodes and smaller, sometimes differently shaped leaves, recommence their elongation. We see again the inhibitory influence of the first leaves of the branch which, as we have shown previously, can lead to the appearance of primitive forms.

Other normally formed winter buds seem to rest during the summer, while they are differentiating embryonic primordia. The period of deepest dormancy appears to be the end of autumn or the beginning of winter. By the middle of the winter, the buds can be made to unfold by simply transferring the twigs into a warm room or, as in the case of the beech (Klebs, 1914), by exposure to strong electric light.[2] A long photoperiod is therefore probably responsible for the development of buds, while the short fall days hasten the advent of the rest period by causing an increasing degree of inhibition. Long days, on the other hand, decrease inhibitions. According to Moshkov (1939), if the photoperiod in the fall is artificially lengthened, experimental woody plants do not show the internal changes which always accompany the rest period. Such plants then have a low resistance to cold and can freeze in a hard winter.

From the point of view of plant integration, it is interesting to note that all the buds do not seem to rest equally deeply. The dormancy of the small buds of the lilac, for example, seems much less deep than that of the large ones toward the top of the plant. It is certainly not just a coincidence that

[2] Or treatment with gibberellin. (Ed.)

these top buds, which directly after their formation enlarge greatly under the influence of the roots, are by the end of the vegetative period most sensitive to the inhibitory influence of the leaves. This demonstrates major shifts in correlative interactions during the course of the vegetative period, which may be altered by applying growth substances. For instance the young, one-year shoots of lilac (Syringa), if smeared with gibberellic acid in lanolin on the upper side of the apical leaves while their axillaries are still very small, show in the following spring an inverse relationship between the large growing and the small dormant buds. The apical buds, in spite of their unusual size, do not grow out until the smaller lower ones have first been excised. It has been made clear that the natural inhibition in plants is increased by the activity of the leaves, and it is well known that this can be artificially intensified, for instance, by spraying auxin solution on the leaves at the time of their greatest activity, in July or August. In this way the flower buds of some fruit trees can be delayed in opening and thus escape damage by late frosts.

The method used for the prevention of budding in stored potatoes rests on the same principle. In this case suitable solutions of maleic hydrazide are applied to the foliage leaves at a given time before the harvest. Gibberellic acid and gibberellins act in the opposite sense, shortening the rest period of the normal winter buds.

When inhibitions in woody plants are not too strong, one can often induce buds to grow out by amputating the leaves. In the same way, amputating the scales will often bring about the unfolding of new bud primordia, even in the presence of leaves on the branch. These responses show that the correlative interdependence of organs determines the arrest of growth for a certain period or forever. The rest period of buds is therefore a correlation phenomenon comparable to the previously described apical dominance.

Apical dominance may provide a possible explanation for

the periodic fertility of fruit trees, which is an especially serious problem in the case of the apple (*Malus sylvestris*). On these trees in fertile (bearing) years only vegetative buds are formed and no flower buds. The overabundance of flowers and fruits appears to inhibit the formation of new flower buds so that in the following spring the tree forms only very few blossoms. The real cause, however, is found in the strong growth inhibition exerted by the leaves. Pea seedlings with the main stem and one cotyledon removed can be considered as models for this phenomenon. The cotyledon which remains on the opposite side inhibits the development of its axillary bud, so that the dominance is transferred to the bud in the axil of the amputated cotyledon. If this bud is cut out in time, the bud of the remaining cotyledon develops freely. Similarly if, by means of a herbicide or simply by hand, we reduce the number of flowers on a fruit tree to one fifth at the time when the buds start to unfold, flower buds can be differentiated for the next year, because at that time conditions favorable to the embryonic development of flowers (the so-called floral stimulus) are present in the plant.

Other methods which are used for the regulation of periodic fruit tree fertility can be understood through experiments with flax seedlings. Flax cotyledons inhibit the buds in their axils only if the root has been removed, while if the root is present the cotyledons stimulate the growth of these buds. The seedlings with intact roots contain a much greater amount of organically bound nitrogen than the others. This correlates with the stimulatory effect of the roots. Since the synthesis of nitrogen compounds, which is localized mainly in the roots, tends to decrease inhibitions, it is recommended to fertilize periodically bearing fruit trees with nitrogenous fertilizers, or at least to stimulate root activity by watering (for example, with manure water). If this is impractical, then the ratio between the crown and the root system can be altered by pruning the former and thus reducing its size.

Periodic bearing in fruit trees is therefore a result of altered correlations between the crown and the roots, and the above-mentioned agrotechnical procedure presumably operates by increasing the relative activity of the roots. This activity usually decreases when crown inhibitions increase, so that sometimes during the summer months roots can become dormant for a short time. In the fall, with colder and rainier weather, the roots grow again and these inhibitions decrease. The previously inhibited buds are thus able to develop also on periodically fertile apple trees, but they are only vegetative.

According to A. A. Nichiporovich (personal communication), periodic bearing of apple trees is an especially serious problem in the Asiatic Republics of the Soviet Union, where sometimes one cannot get certain types of apples for 2 years. This behavior is not due to the exhaustion of reserve substances by an overproduction of fruits, because an analysis of the reserve content does not show changes from one year to the next. It must depend upon very sensitive correlations between roots, leaves, flowers, and fruits on the one hand, and the buds which are to be initiated for the following vegetative period, on the other. The problem of plant integration thus becomes very obvious.

It is quite possible that in the case of deciduous forest trees the exhaustion of reserve substances in fertile years might be the main cause, for the fruits do use up a great deal of the supply for their development. According to Gäumann (1935) the beech uses up practically all of its carbohydrates when its flowers are formed, and it takes sometimes more than ten years before the level of carbohydrates can be restored to such an extent that new flowers and beechnuts can develop. On the other hand, the level of nitrogenous organic substances remains relatively constant. We see again the special place of nitrogenous substances in the life of the plant, which is stimulated by them in a very economical way.

The initiation of flowers requires specific conditions which depend on the plant's adaptation to its environment, as is explained in Lysenko's theory of phasic development; this theory particularly stresses two phases: seasonal and photosensitive. The first concerns annual and biennial plants, the second those plants which are adapted to various photoperiods, depending on the latitude. In central Europe during the summer growing period the days are long (16 hours or more), but in the tropics they are always short (close to 12 hours).

The role of specific substances in the formation of flowers is not yet satisfactorily clarified. In our experiments we first showed that auxin and other substances, similar in effect, usually inhibit the formation and development of flowers, since auxin paste applied to the buds on isolated segments of *Circaea intermedia*, which would normally have formed floral shoots, caused them to change into runners and tubers. The same transformation was achieved as when the bud had been immersed in water or placed in an atmosphere containing ethylene. A well-known exception is the pineapple (*Ananas sativus*), which can be made to flower under the influence of a very small amount of ethylene or a synthetic auxin, or by a change in production of natural auxin, which happens if the plants are laid horizontally. Perhaps there is here the basis for an explanation of the increased production of flowers and fruits on horizontally trained branches of fruit trees.

An antagonist of auxin, 2,3,5-triiodobenzoic acid, acts as a stimulant of flower and fruit production in certain plants. For instance, it increases the percentage of wheat shoots which flower before the end of the normal vegetative spring phase (Šebánek, 1955).[3] This action may be the consequence

[3] The classical instances are the production of flowers on very young seedlings of tomato (J. deWaard and J. W. Roodenburg, *Koninkl. Ned. Akad. Wetenschap, Proc.*, Amsterdam 51:248–251, 1948), or in the axillary buds of tomato or soybean plants (P. W. Zimmermann and A. E. Hitchcock, *Contrib. Boyce Thompson Inst.* 12:491–496, 1942); A. W. Galston, *Am. J. Botany* 34: 356–360 (1947). (Ed.)

of its inhibitory effects, which are seen in other plants in the modification of the vegetative tips and apical meristems, in such a way that internodes fail to develop between the youngest leaf primordia and leaves grow together into cup-shaped structures. As a result, triiodobenzoic acid does not stimulate flowering in those plants.

Some growth inhibition is indeed necessary for the formation of flowers, for vegetative growth has to slow down. One can thus explain the action of maleic hydrazide (used widely where growth is to be inhibited) in increasing the number of fertile cobs of corn when the young seedlings have been treated with a dilute solution of this growth inhibitor (Cílková, 1957). The solutions, however, have to be weak enough not to decrease the vegetative growth so as to let the female cobs develop normally.

The reason why gibberellic acid does not have this effect on corn is probably because flowers are formed on the corn plant, as in other cereals, at the time when the internodes between the meristematic nodes have not yet elongated. This substance is involved in the formation of flowers only when it is required for the elongation growth of the axis. Gibberellin can, however, be substituted for the low temperature effect in some cases. An example is given by the biennial strain of Hyoscyamus, in which one-year-old leaf rosettes, even without a temperature drop, are induced to form a long floral shoot by application of gibberellin. Scrophularia also requires low temperature; if its tubers are kept over the winter at room temperature, they form in the light only a tight rosette with large leaves; however, the application of gibberellin to such plants again produces a long shoot with an inflorescence. The same is true for the short-day rosette plants of *Bryophyllum crenatum*, which can grow for years without forming any appreciable internodes and with round leaves borne close to the ground, if they receive no special care during the winter and in the summer are exposed to light for only 8 to 9 hours a day. If the plants are treated with gibberellin and kept on short days the primordia of the upper internodes elongate and flowers develop at the tip of the stems. But if after the same treatment they are exposed to long days, even though the internodes elongate no flowers are produced. Similarly, gibberellin has no effect on Bryophyllum if from the beginning it is grown under long-day conditions, because in that case it forms on its own account long internodes and oval petiolated leaves.

Bryophyllum crenatum, as also *B. verticillatum* and *B. daigremontianum*, flower only if provided with alternating long and short days. Individuals grown in long days have to be kept at least 10 days under short-day conditions in order to acquire the ability to form flowers both in long and short days, while individuals cultured in short days have to be kept at least 20 days under long-day conditions and then at least 10 days under short days, subsequently, in order to develop floral induction.

Of these various photoperiodic cycles, the long-day phase (during which the leaf rosettes form long stems) can be replaced by gibberellin treatment in plants grown under short-day conditions, but the plants so stimulated then have to be kept in short days in order to flower. Thus, gibberellin cannot replace the short day even in these "long-short day" plants. The same is true for all short-day plants, as has often been published. Gibberellin is ineffective for these plants because in long days they elongate even without gibberellin, though they do not flower. This explains why the above-mentioned species of Bryophyllum flower fully in our country at the beginning of winter, that is, at the end of December and beginning of January. They start to form flower buds only after the fall equinox, when the days have become shorter than 12 hours. This is probably their upper threshold for short days, while 13 hours would constitute long days. Such a narrow range probably represents the conditions of the equatorial homeland of these plants, natives of Madagascar, and is only one of many properties adapted to homeland conditions which have been maintained in culture. Another such property is the production of aerial roots everywhere there is a blockade of auxin, as in the nodes and curves of the stem and the undersides of axillary branches. It is notable that these plants do not react visibly to the great differences in temperature characteristic of our climate, so long as the temperatures do not fall below zero. Also these plants show no evidence of a requirement for a temporary cold spell for the formation of flowers, because they have made no such adaptation; for decades they have been cultured in our climate without any hereditary changes.

Lysenko (1954) and many other workers have shown that the transformation from the vegetative to the reproductive stage, which is brought about in winter cereals and biennial plants by brief exposure to cold, results from a direct effect on the vegetative tips; this is in contrast to the transformation caused by photoperiod, which depends on the duration of illumination of the leaves, from which the signal is then transmitted to the vegetative tip (Figs. 36 and 38). This explains why, if the tip of the plant is kept at a warm temperature and only the rest of the plant is cooled, there is no transformation of the vegetative into the reproductive state.

The apical meristems take up organic substances, especially sugars, from the mature parts and reserve organs, but they are not passively influenced by these substances. Their reaction to cold is that to which they have become adapted during their phylogenetic development, and it can be changed only by altering the process used to bring about transformation from the vegetative to the reproductive state; this, of course, requires a changed heredity (Stoletov, 1957). In the case of annual rhythmicity we also observe a certain independence of individual buds from one another. For example, the effect of ethylene chlorhydrin can be limited to a single bud, which is forced to grow on an otherwise dormant branch of lilac, while the opposite bud may remain undisturbed.

In this book we have given many examples of the influence of nutrient substances, originating in the mature parts, in

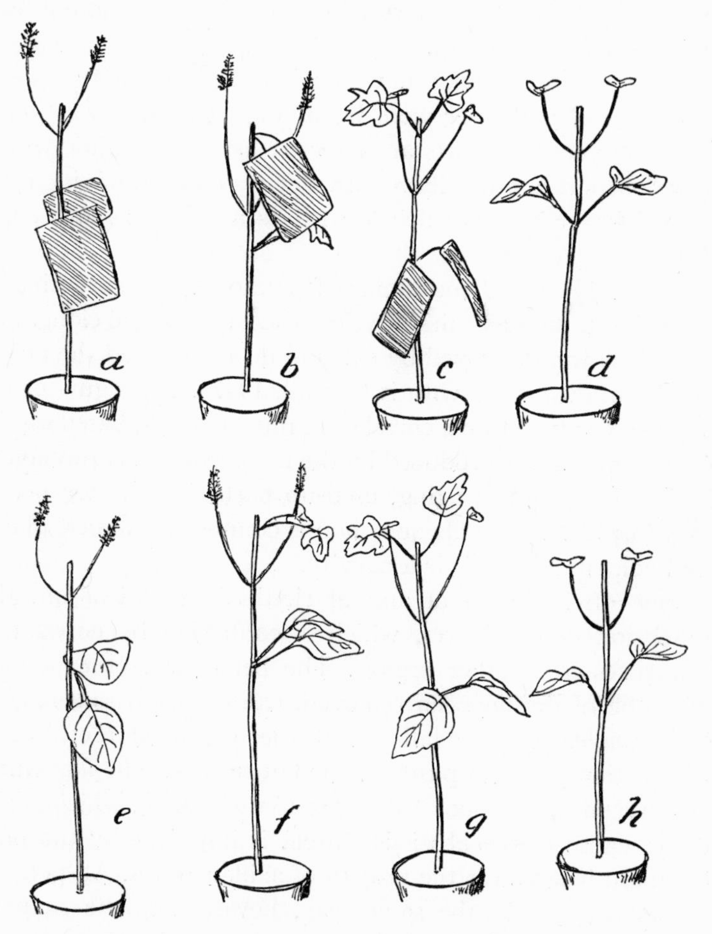

Fig. 38. *Perilla nankinensis:* (*a*) leaves in short day (covered with black paper for part of each day), flowering occurs; (*b*) only upper leaves in short day, flowering also takes place; (*c*) lower leaves in short day, plant vegetative; (*d*) leaves in long day, plant vegetative; (*e* to *h*) the same plants without the black paper covers, to show the leaves (after Chailakhian).

promoting the development of the primordia of buds during embryonic development or during active elongation. Conversely, growing buds inhibit the growth of other primordia, especially when the growing leaves or flowers are of a certain minimal size; quite small rudiments or meristems do not have much inhibitory effect (Snow, 1929). Thus, in the experiments with Scrophularia, which had been defoliated when the buds of the tuber were beginning growth, it was noted that the youngest parts are affected by the inhibitory influence of more mature leaves. In this way plants having sympodial branching can develop leaves and fruits more effectively.

Similarly, apical meristems of potato tubers, sprouting in weak light and low humidity, tend to die off. The elongation zone is slowly destroyed by calcium deficiency and the apical meristem dies along with it (Dostál, 1943). The aging of the meristem is therefore a correlation phenomenon, based on the strong inhibitions produced by the mature parts. A physiological regeneration of aging, used-up parts, is therefore necessary in plants. This phenomenon is common in woody plants and contributes to their longevity.

One obvious sign of annual rhythmicity is leaf-fall, the abscission of the leaves, which is controlled by correlative influences from other organs. The behavior of the petiole, after the blade has been removed, can be taken as a model. The petioles normally fall off, but they can be kept on by the application of auxin paste on the cut surface. The formation of the separation zone at the base of the petiole, which is the direct cause of the abscission itself, is inhibited by the flow of auxin, just as was the growth of axillary buds. All petioles do not behave in the same way, however. If the stem of *Bryophyllum crenatum* is cut up into segments possessing one node each, from which the leaf blades are removed, and if auxin paste is then applied to the stub of one petiole and water paste to the other, we observe that the petiole treated with auxin is retained longer only if the segments are taken from the base of the stem. On segments originating from the upper part of the plant the auxin-treated petiole falls off even sooner than the control. A difference in the transporting activity of the petiole in different parts of the plant may be involved here, for petioles of mature leaves conduct assimilatory products out of the leaves from the tip toward the base, while petioles of young growing leaves bring in nutrient substances in the opposite direction, from the base toward the tip.[4]

The occurrence of other leaves or of roots on the plant or the regeneration of roots on a segment are very important factors in the life of the petiole. The presence of leaves hastens the abscission of the petiole, while regenerating roots keep the petiole on the axis for a longer time. In *Circaea intermedia* plants the removal of both buds and stolons prevents the abscission of debladed petioles (Dostál, 1911). It is particularly in woody plants that the dependence of petiole abscission is controlled by the integrity of the whole plant, since this process is considerably delayed by an active root system. When *Acer negundo* is cut down, sprouts grow out from the large stumps. If these are debladed at one side of their nodes, they retain the uppermost petiole for a very long time; the lower petioles, however, drop, and in consequence they develop lateral shoots. Debladed horse chestnut (*Aesculus hippocastanum*) plants abscise their petioles in order toward the base, the upper (although the youngest) sooner than the lower ones, which are the largest; this parallels the inhibitions which, present already in the winter buds, increase acropetally. But when the strongest branches of tall trees were debladed on June 14th, they retained their petioles until the end of October. This may be explained as a correlative compensation for their lost leaf blades, induced by the direct connection of the apical parts of the branches with the roots.

[4] The explanation may also be that the young leaves more readily produce ethylene in response to auxin, and the ethylene then causes the abscission. (Ed.)

According to this the fall of petioles took place in acropetal order, as is usual in these plants.

Ethylene acts as an antagonist of auxin in hastening the abscission of leaves. In an ethylene-containing atmosphere, large, whitish calluses arise under the needles of the yew tree which set loose not only the needles themselves but even stronger twigs, which then fall off. If those needles are isolated, joined to a very small piece of axis, there forms between these two parts a voluminous callus; in pure air the callus forms only over the scar of the axis, and the needle remains firmly attached to the stem.

Leaf abscission is also accelerated by the growth of young leaves and flowers, in deciduous woody plants and in evergreens. External factors, such as strong light and insufficient humidity, also increase the rate of abscission, that is, they act in the same sense as do younger parts of the plant

In woody plants the development of leaves is coupled to the formation of annual rings, which arise under the influence of auxins originating in the developing buds. Jost (1891) found long ago in his experiments with bean seedlings that if one primary leaf is amputated early enough, the development of vascular bundles is markedly decreased on that side.

Later, Snow (1933) showed a close relationship between the production of auxin in growing leaves and the activity of the cambium. It is easy to follow the progressive spreading of some influence from the developing buds to the cambium, which then undergoes renewed activity as a result. Especially in trees with ring-porous wood and very wide vessels in the spring wood, rapid formation of spring wood vessels spreads from the buds down to the roots. There is, however, some evidence that similar impulses may originate in the roots, independently of the activity of buds up above. In both instances there may perhaps be formation of auxin from tryptophane, during the formation of new proteins. It was in fact possible to force experimental plants, which had been kept in a warm greenhouse since before the beginning of winter and therefore had neither buds nor rings, to form new wood under the influence of auxin applied to the upper section. Dormant plants have first to be activated by ethylene chlorhydrin.

The differentiation of the woody tissues into vessels, fibers, and parenchymatous cells is not primarily the work of auxin, but rather that of special reactions which are not yet clarified.[5] The formation of the spring wood, due to the inflow of auxin from the opening buds and to the auxin liberated directly into the active cambium from reserve substances, is followed by the formation of summer wood, characterized by much thicker vessel walls and fibers, due to the influence of the photosynthetic products of mature leaves; these leaves can sometimes totally inhibit the formation of wood, so that the products of assimilation accumulate instead as reserve substances in the roots, trunks, or leaves. These same inhibitions also stop all other growth, with the exception of the embryonic differentiation of winter buds made possible by the influence of bud scales. The cambial activity can be renewed after the loss of leaves, providing the bud primordia can still develop in the same year. If, however, the two types of wood are not formed under the influence of the growing shoots, it is a sign that the cambium has remained active continuously from April to June, or has not changed its activity enough to form summer wood transitionally. Obviously the differentiation of buds, leading to the formation of summer shoots, is not linked to a real rest period, which comes only much later.

Great differences exist among the trees in temperate climates, and even more so among the exotic woody plants adapted to various tropical climates. But in general there seems to be a close correlation between the development of

[5] There is good evidence that an interaction between auxin and other factors controls the formation of normal wood; for auxin and kinetin see Sorokin et al. (1962); for auxin and gibberellin see Wareing et al. (1964). (Ed.)

leaves and that of the wood, involving a regular periodism. Seasonal rings can be simulated by certain models of periodicity, such as Liesegang's rings; if we put on a layer of gelatin containing potassium bichromate, a crystal of silver nitrate, concentric red circles of silver chromate appear, separated by circles of yellow gelatin. This is, however, only a very vague resemblance, quite insufficient to represent the complicated correlation conditions existing between organs and tissues of plants.

CHAPTER VIII

External Factors and Integration

Michurinian biology rightly stresses everywhere the unity of the organism with its environment. The phenomena of integration provide the best example of this law. There is no definite borderline between external influences and the internal factors which have been discussed in the preceding chapters, since correlations, which are the basis of integration, are the result of the action of external factors on the living matter. Indeed, correlations are rooted in external factors from their very beginning, as we realize when we follow the embryonic differentiation both of animals and of plants. Structures which will be used only much later by the plant for its physiological needs are already present in the embryonic stage. It is readily understandable that the same external factors which act at the origin of correlations can eventually bring about great changes in them; in this resides the possibility of the transformation of plants. This problem was considered in connection with polarity and rhythmicity, both of which depend on correlations. Experiments on regeneration and transplantation lead us to the same conclusion; the most important factors, on which the life of the plant directly depends, act not only during its visible growth but also during its embryonic origin. These factors determine (usually indirectly through the mediation of mature parts) the shapes and anatomical structures of the primordia, closed off from the outside world by bud scales or by fruit or seed envelopes.

Light, for example, does not have to fall directly upon the differentiating primordia, in order to produce shade-tolerant or shade-intolerant leaves; which of these two types is produced is probably decided by the products of metabolism in the leaves of the plant during the previous year. These products would have been different in the leaves on the strongly lighted peripheral branches from those on the shaded ones in the center of the crown.

Of basic interest are the experiments which originally led to the discovery of the main regulatory substance of plants, namely auxin. Darwin (1880) covered the tip of the seedling of Coix with an opaque cap and illuminated the plant from one side. While uncovered plants turned toward the light, plants with covered tips remained straight. Much later Went (1926) discovered in the isolated tip, set on a block of agar, a growth substance which could be shown to promote growth of the seedling. Subsequently, after a number of errors, this substance, found also in urine and as a product of the fungus Rhizopus, was identified both in Holland and in the United States as "heteroauxin," or indoleacetic acid. Evidently it flows from the tip to the lower part of the coleoptile, where it initiates elongation growth, while the tip itself does not grow. The bending is due to the fact that the shaded side contains more auxin, and therefore elongates more rapidly than the side toward the light.

Němec (1900) found in the tip of roots and of the coleoptiles statolithic starch, which he believed explained the sensitivity of these parts of the plant to gravity. Shortly afterwards he discovered, in experiments which were described earlier (Chapter IV), that root tips can be regenerated after having been injured, if the wound is less than 1 mm from the meristem. The regeneration takes place through the joint action of very unusual influences which do not depend merely on nutrition. He thus gave a major impetus to the attempt to explain correlations, which are most obvious in just such regeneration processes. All of the interactions between individual parts of the plant depend to a greater or lesser extent on the influences of external factors. From the experiments of these two workers stem many of the current explanations of correlative phenomena.

Light becomes influential immediately after the establishment of the most important of the plant's correlations, namely that between the stem and the root. This occurs even in the zygotes of algae, where, after only a few hours and under the influence of a unilateral light source, a definite separation is established between the photosynthetic part and the rhizoids. In higher plants also, polarity is modified by light and always remains dependent on it to a large extent.

Another important correlation, that of apical dominance, is weakened by strong light. In other words, as in the original experiment of Charles Darwin, the normal inhibition is impaired by light. This can be demonstrated as follows: pea seedlings are exposed in a closed vessel to vapors of the methyl ester of naphthaleneacetic acid, which inhibits their growth. After 4 days the plants are taken out of the chamber; they now grow normally if exposed to light, but remain permanently inhibited in the dark. The same interpretation can be given to the effect of light on the formation and growth of buds, which it stimulates, while on the contrary roots grow better on the less illuminated side. Another example is provided by the non-cellular alga *Caulerpa prolifera;* on the side of the rootstock exposed to light, initials of leaflike assimilators and new rootstocks appear, while on the shaded side there grow only rhizoids. The quality of light is important also; leaf primordia are formed in blue light while in red light they are inhibited.

The different conditions favoring the initiation of roots and leaves do not involve merely morphology. There may be an exchange of function between them, for if the roots are exposed to light and the leaves kept in the dark, then the roots form chlorophyll and carbohydrates, while the leaves in the dark can produce amino acids (which is always true for very young leaves). External conditions can therefore radically change the metabolism of organs, even to the point of counteracting their original functions.

Light has a great deal of influence on the correlations between the underground and the aerial system. In weak light, with slow photosynthesis, the resulting carbohydrate deficiency brings about an increase in the ratio of the dry weight to the stem to that of the root. Thus plants in the shade have their stems better developed than their roots, while in strong light (as is shown by plants on high mountains) the stem:root ratio is decreased.

Day length is an important factor, because for long-day plants the stem:root ratio is higher in long days, while for short-day plants it is higher in short days. Day length is also, of course, related to the production of flowers and tubers.

Temperature influences some aspects of plant differentiation, as is shown particularly by tuber-forming plants. In Vöchting's experiments (1900) with potato tubers of the Marjolin variety, daughter tubers were formed directly on the mother tuber at 6° to 7°C, while at 25°C branching shoots were formed first. Tubers of *Circaea intermedia*, which have not completed their rest period, continue to grow at room temperature in the same form, with thick short internodes and scale-like leaves. In these cases low temperature acts as a major stimulant. It can be replaced sometimes, as in the case of simple elongation, by gibberellin, but dormant buds of Ficaria or lilac, for example, cannot be made to grow by applying this substance. On the other hand, higher temperatures increase the ratio of aerial to underground organs.

Plants forced to grow prematurely, before the completion

of their rest period, often show traces of primitive forms of growth. A striking example is given by the so-called "frost atavisms." Frost-bitten plants, which regenerate flowers from incompletely differentiated primordia, can manifest suggestive ancestral forms; the flowers of the convolvulus, for example, may have separate petals instead of the normal, bell-shaped corolla with fused petals.

An equally important factor is water supply. Its deficiency decreases the stem:root ratio, while an excess of water increases it. The aeration of the soil may also be involved, since a moderate soil humidity, which is not high enough to interfere with the respiration of the root system, speeds up metabolic turnover and thus contributes also to the stimulation of aerial parts. Krenke, in his theory of cyclic aging and rejuvenation, cites the example of the sugar beet, which clearly shows the influence of age on the shape and size of the leaves. Each leaf shape was found to correspond to a definite sugar content in the root. The shape of the root, also, could be markedly changed by culture conditions, so that a shape characteristic of older plants developed much earlier when the water supply was low. On the other hand, an excess of water contributes to the retention of juvenile leaf forms; on a plant which was kept dry, the 13th leaf showed a shape as advanced as the 19th leaf on a watered plant. The physiological and the chronological ages thus do not always exactly overlap. Weak light and high humidity act in the same direction, both on those correlations which control the shapes of plants (including, to a certain extent, the degree of etiolation) and on those correlations which determine the plant structure.

Nutrition can also modify form. The roles of nitrogen and zinc are particularly important in the regulation of growth of green plants; both elements are also probably involved in the synthesis and action of auxin. The correlations between the stem and the root system depend therefore to a large extent on the available nitrogen in the soil. If the nitrogen supply is limiting, most of it is used up directly in the root for the synthesis of amino acids. At the same time carbohydrates produced in the stem are also used up. Only a very small amount of amino acids reaches the stem, so that the stem:root ratio decreases. On the other hand if there is an excess of nitrogenous compounds, the roots use up only a small fraction of them, and most of the amino acids flow to the leaves to form new cytoplasm. A larger quantity of carbohydrate is used up in this case, and hence only a little remains at the disposition of the roots, which then grow less and the stem:root ratio increases. The level of nitrogen in the soil can be raised by the soil microflora, many members of which fix atmospheric nitrogen. This property accounts for the propitious influence of *Azotobacter chroococcum*, the most common nitrogen-fixing microorganism, on the aerial development of plants.

Correlations between leaves and buds are also sensitive to the mineral supply. Thus, mutual inhibition could be established between the two opposite buds on an isolated segment of the stem of Scrophularia, carrying two leaves, by making use of the fact that nitrogenous salts decrease bud inhibition, while phosphates increase it. Nitrate was supplied to one side and phosphate to the other. The shoot on the side supplied with nitrate became dominant over the shoot on the side supplied with phosphate. Among the microelements, zinc has a most important role in correlations and especially in the control of the dry weight stem:root ratio. When potato seedlings were grown in water, they showed a remarkable increase in the stem:root ratio if zinc sulfate was added to the culture; the value was 8.68, as compared to 4.39 in plants supplied with Crone's nutrient solution. Plants in the complete nutrient medium A-Z, with addition of trace elements according to D. R. Hoagland, gave a ratio of 6.45. On the other hand, the dry weight of tubers produced was 0.35 gm per plant in a zinc-enriched solution, 1.53 gm with the further addition of borate, and 1.75 gm in the complete solution.

Fig. 39. Potato (*Solanum tuberosum*): (*a*) previously wrinkled tuber with large apical sprout, treated at base with maleic hydrazide, absorbs the solution but forms no tubers; (*b*) a similar tuber which has absorbed triiodobenzoic acid forms a highly branched sprout with many tubers.

Synthetic substances of the auxin type also have an inhibitory effect. They are not usually found in plants, and the only characteristic they have in common with animal hormones is that they act in minute concentrations.[1] They can be used to modify normal correlations and are therefore valuable tools for the clarification of interactions between plant organs. When they are used as growth factors, especially in the treatment of seeds prior to sowing, great caution is necessary, because the proportions found effective in the laboratory can be greatly modified out of doors. Thus auxin (indole-3-acetic acid) which acts in plants as a natural

[1] And they have precisely the same biological effects as indoleacetic acid, with the exception of transportability from apex to base. (Ed.)

growth factor, has many different effects if applied externally, depending on the state of the plant, its age, the types of tissues present, and so on. Auxin therefore lacks the specificity of action characteristic of animal hormones.

The same can be said about triiodobenzoic acid, which in higher concentrations limits apical dominance, stimulates branching, and decreases vegetative growth. It is thus capable of stimulating plant productivity, when this involves production of tubers or of seeds and fruits. An example is offered by a potato tuber to which triiodobenzoic acid is applied by means of a wick of cotton tied around the base of the sprout. Even a very withered tuber regains its turgor, the sprouts start to grow, branch out, and soon form a large number of tubers (Fig. 39b).

Gibberellic acid and gibberellins influence especially the correlations between the axis and the leaves, so that the axis elongates abnormally at the expense of the leaves. Ultimately the leaves may lose their chlorophyll synthesizing power. No more than 10^{-6} grams of gibberellic acid can transform one of the dwarf varieties of corn into a normal plant. There is no doubt that differentiation processes result from the action of specific substances, but their nature is still not clarified.

Overgrowths such as tumors and galls (hypertrophies) represent still another unsolved problem, especially those with a complete organization of tissues which are not normally present in the plant, and which are caused by special substances diffusing from the parasite. Only some of the simpler forms, such as certain galls, can be explained by an interference with normal correlations, as, for example, the previously mentioned monopodial branching caused by *Eriophyes loewi* on the lilac. Other cases, however, are caused by viruses or other factors.[2]

The influence of gravity is crucial for plants. By directing

[2] For an analysis of the witch's broom caused by a bacterium, and traced to its synthesis of a cytokinin, see Sachs and Thimann (1966). (Ed.)

the flow of natural auxin downward, it helps in the formation of the root on the lower pole and of the buds at the upper pole.[3] Horizontal branches or stems behave as though the flow of auxin toward the base is slowed down in them, so that they flower more and grow less. On a horizontally laid stem of Scrophularia, at the close of flowering, little side branches grow from the upper buds all along the stem. Among these branchlets, those closest to the tip bear flowers, those in the middle bear leaves, and those near the base turn into tubers. This shows that apical dominance has been partially suppressed along the upper side.[4] Similarly all along the trunk of an uprooted coniferous tree, lying horizontally, the branches of the upper side grow vertically upward, unaffected by the apex of the tree which also turns upwards. Segments of willow branches, hung up in the inverse position, form roots further away from the basal pole and buds further away from the apical pole than when in the normal position. Gravity also acts on the alga *Caulerpa prolifera*, since if we turn it through 180°, it forms leafy assimilators on the originally basal side (now at the top) and rhizoids on the morphologically upper side (now at the bottom). All of this shows, therefore, the great influence of gravity on the flow of auxin, which is the most important of the known activators of the differentiation processes.

Mechanical influences can also alter correlations, almost certainly because they affect the distribution of auxin in the plant. The mechanical structure of the plant is basically determined in the embryo and cannot be changed to any substantial degree during growth by pressure, pulling, or bending (Razdorsky, 1955). In spite of this, plants are sensi-

[3] This statement is remarkable because, until the work of Hertel and Leopold (1963), no evidence for an effect of gravity on the *longitudinal* transport had been given. (Ed.)

[4] Growth of axillary buds along the upper site of horizontal stems is often seen in wild plants, e.g., leafy spurge (*Euphorbia esula*) and the New England aster (*A. novae-angliae*); cf. Delisle (1937). (Ed.)

tive to such factors, and we can make one of the buds on a node of a hop plant grow by repeatedly touching it, while the other bud will not develop. On bowed stems or roots, lateral branches grow more readily from the convex side than from the concave. This fact can be used in fruit growing to decrease vegetative growth and increase fertility. Here again it is a matter of the distribution of auxin, and there is no need to speculate about any mysterious conscious sensitivity of plants as indicated in the "morphestesia" of Noll (1900).

It is also possible to explain in terms of the movement of growth substances the well-known fact that roots grow from stems or from the main root directly outwards and spread, if they are strong enough, radially into the environment. Runners behave similarly, bringing about the spreading of the plant and its vegetative multiplication, especially when their tips enlarge into tubers which accumulate reserve substances. As long ago as 1892, Sachs was interested in the question of why in a pot the roots and runners are packed against the sides. A simple experiment has demonstrated to the author that this is due to a differential distribution of auxin in these organs, when they meet an obstacle. An auxin solution was applied from time to time to one side of a porous pot in which *Circaea intermedia* was growing, while the other side was treated with pure water. When after a time the pot was emptied it could be seen that the runners which formed a continuous belt at a certain depth in the soil turned their tips, like little hooks, away from the soil on the side treated with water, while on the side treated with auxin they remained stuck to the soil, as they had been in the pot (Fig. 40a). In the latter case the outer side of the growth zone of the runners received the more auxin, which equalized the difference between the auxin contents of its convex and concave side. On the water treated side of the pot, some influence, perhaps the pressure of the pot itself, caused an increased concentration of auxin on the concave side of the runner. It is easy to imagine that a runner, and similarly also a root, can thus circumvent any obstacle in the soil and continue in its original direction.

The same plant demonstrates very simply the influence of the depth at which the plant is set in the soil, a very important factor in plant development and one which can influence the productivity of agricultural plants. If the runner of *Circaea* is kept in the dark throughout the experiment it continues to grow horizontally (Fig. 40b). If, however, we illuminate its basal and middle part (which are not growing), then the tip, which is relatively far from the light, bends in the dark

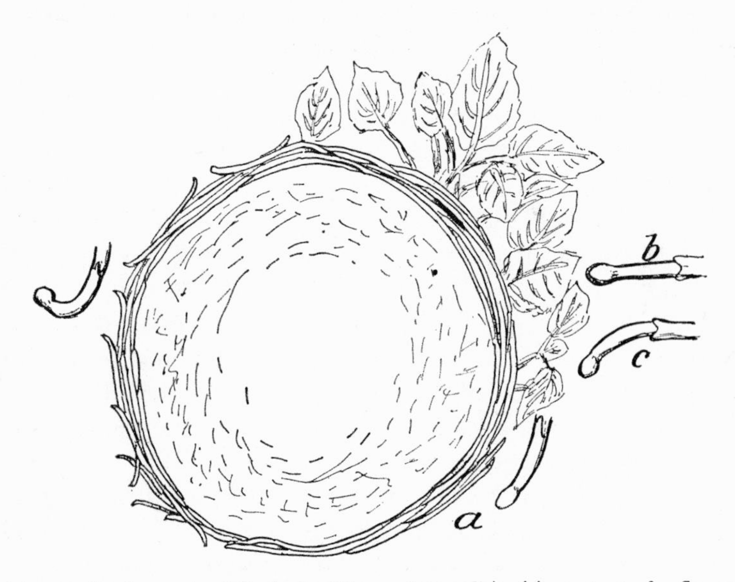

Fig. 40. Enchanter's nightshade (*Circaea intermedia*): (*a*) content of a flower pot, the right side of which had been treated with auxin solution and the left side with water; after taking out of the pot the tips of the runners at the right remain stuck to the soil, while on the opposite side they curve away from the soil and shorten, especially when moist; (*b*) tip of a runner which was in darkness in the soil continues to grow straight; (*c*) tip of a runner, the basal and middle parts of which were exposed to light, turns downward (depth regulation).

obliquely downward (Fig. 40c). The depth of runners and tubers is therefore determined by the aerial parts of the plant, exposed to light. This is evidence for the harmonic union between the aerial and the underground parts of the plant; the illuminated parts of the plant bring about a certain change in the tip of the runner which shows up as a transformation of neutral geotropism into a positive response.

There is a relationship here with the previously described experiments on isolated segments from the base of the stem of *Circaea intermedia*, which usually form horizontal runners, turning down into the substrate at their tips. Their plagiotropism — or, in the light, their positive geotropism — can be transformed into a negative geotropism by applying a small amount of gibberellic acid paste, giving rise to vertical stems with long internodes (cf. Fig. 28e).

Conclusion

The results of the experiments presented here justify the conclusion that in plants one rarely finds a regulatory mechanism comparable to the neurohormonal system of animals. Plants lack both nerves and endocrine glands producing specific hormones, transported to the rest of the body by the circulation of blood and lymph. They are nevertheless just as alive as animals; however, the more active expressions of life, such as motility, are rendered impossible by the thick cell walls, which, in trees, for example, make up 90 percent of the dry weight of the plant.

Plants evolved through phylogeny in a very different way from animals, if only to arrive at the particular nutritional mechanism they possess, namely assimilation of mineral substances. This is essential not only to their own life, but constitutes the energetic and material basis of all other life on earth. They therefore developed roots and leaves which can be renewed continuously as they become used up or lost. Hence even the most highly organized plants, either those which are phylogenetically the most recent, such as the cereals, or those which are structurally the most complicated, such as the trees, show no traces of a centralization comparable to that in animals, for it would give them no advantage.

Darwin (1881) must have written only in a figurative way when he saw in the tips of the coleoptiles of grasses the brains of plants. We know today that the apex is the center of activation of the most important regulatory substance of plants: auxin (indoleacetic acid). Auxins can, however, originate in many other parts of the plant if protoplasm is being synthesized there from substances supplied by other organs, such as cotyledons, endosperm, tubers, et cetera. It is for this reason that leaves and reserve organs are the main regulating centers of the plant, but there are so many of them that it is quite impossible to recognize any one main regulation center which would represent the overall control of plant integration.

Leaves in the light produce many other substances, which control plant differentiation and accumulate in reserve tissues. The nature of these substances, which are probably as varied as the differentiation processes themselves, is not yet fully clarified. Among them, auxin, gibberellin, kinin, and other native substances found in plants are activators of specific processes, just as are certain synthetic substances with interesting effects, such as triiodobenzoic acid.

A great step forward in the understanding of morphogenesis came when Němec (1905) proved, against the then prevalent notion of the exclusive importance of general nutrients, the specificity of some of these special influences, and showed that they act as early in the plant's life as the embryonic foundation of the new apical meristems.

The study of the integrity of the plant is first of all a problem of experimental morphology. The results of our experiments support the words of Timiriazev (1951) that "everything here is realized with an iron necessity." For example, the correlation which is expressed by the inhibition of the bud in the axil of the remaining pea cotyledon, giving the dominance to the bud in the axil of the opposite amputated cotyledon, is a rigid enough law to convince us that the interactions between the different parts of the plant, on which rests the plant's integration, have a material basis. The cotyledon, through its inhibitory effect, determines which one of the bud primordia (both of which are equally capable of growth) will grow the more under a given gradient of reserve substances; as a result, that one primordium becomes the center of auxin production and later is transformed into an inhibiting organ which will arrest the growth of the other primordia and attract to itself all of the available plastic material.

Under normal conditions, the control of plant integration lies in the limitation of the *number* of primordia allowed to develop. Should they all develop without any regulation, then instead of the harmonious organization which leads each individual through an ordered succession of events, from the embryo to growth, maturity, reproduction, and renewal in the next generation, there would be chaos.

Another experimental object, the flax seedling, demonstrates the important role of roots in the regulation of plant organization; they give to the leafy cotyledon a stimulatory power. The action of the roots is due to their nitrogen metabolism, while on the other hand, the inhibitions exerted by the leaves arise from their predominantly carbohydrate metabolism. The actual nature of these stimulations and inhibitions is not yet fully understood, and for the moment we just have to accept the fact that they are established very early in the embryo, long before their effects become visible on elongation growth and on the shape of the organs.

The concept of the evolution of living things, which was clarified in general by Darwin and stressed for plants by Timiriazev, when he proclaimed morphology to be the basis of the study of evolution, is that plants are the product of a long series of evolutionary processes, which are repeated in shorter form during their ontogenetic development. Interference with this law sometimes brings out the nature of the primitive forms which have preceded the recent ones on a plant. These primitive forms may resemble the common ancestors of today's plants. They also show remarkable concordance with the systematic classification of plants, although this rests on completely different methods.

Since the correlations which exist even in very small plant primordia correspond probably to past ontogenesis, which was included in the phylogeny and can probably be isolated from it, the possibility is offered of a glimpse into the laws of phylogeny and at the same time a perspective of how evolution could be hastened through similar methods.

Much more approachable are the correlations which are at work during the visible elongation of organs. Any new discovery clarifying the laws which govern their interactions must have great practical value for the growing of plants. We must respect the complex harmony of correlations which determine plant integration, and keep in mind also the basic Michurin law of the unity of the organism with its environment — the fact that each external interference with the life of a plant, however localized, influences the plant as a whole.

Bibliography

Index

Bibliography

Ashby, G. 1950. "Studies in the morphogenesis of leaves. VI. Some effects of length of day upon the leaf shape in *Ipomaea coerulea*." *New Phytologist* 49: 375–387.

Blaringhem, L. 1911. *Les transformations brusques des êtres vivants*. Paris: E. Flammarion.

Bykov, K. M. 1950. "Razvitije idei I. P. Pavlova." *Naučnaja sessija I. P. Pavlova. Akad. Nauk SSSR* 513–518.

Cavara, F. 1923. "Fatti di correlazione ed ormoni nelle piante." *Boll. Orto Botan. univ. Napoli* 7: 265–276.

Chailakhian, M. Ch. 1955. "Celostnost organizma v rastitelnom mire." Erevan: *Izv. Arm. SSR.*

——— 1956. "Ontogenes i celostnost rastitelnogo organizma." *Botan. Zh.* 41: 491–509.

——— 1958. "Osnovnye zakonomĕrnosti ontogeneza vysšich rastĕnij." Moscow: *Izv. Akad. Nauk SSSR* 1–78.

——— and T. V. Nekrasova. 1956. "Vlijanie vitaminov na preodolenie poljarnosti u čerenkov limona." *Dokl. Akad. Nauk SSSR* 111: 482–486.

Cholodny, N. G. 1939. "*Fitogormony.*" Kiev: Izv. Akad. Nauk SSSR.

Chtuk, A. A. 1946. "Biochemičeskije izmĕnĕnije privitych rastĕnij." *Usp. Sovrem. Biol.* 21: 357–381.

——— D. Kostov, and A. Borodinova. 1939. "Alteration in the alkaloid composition due to the influence of stock upon scion in *Nicotiana. Dokl. Akad. Nauk SSSR* 25: 477–480.

Cílková, M. 1957. "Stimulační pokusy se sudankou, čumizou a čirokem." *Sb. Česk. Akad. Zeměděl. Věd. Rostlinná Výroba* 3: 430–440.

Daniel, L. 1921. "A propos des greffes de Soleil sur Topinambour." *Compt. Rend. Acad. Sci.* Paris, 172: 612–614.

Danilevskii, V. N. 1913. *Fiziologia čeloveka.* Moscow, I.

Darwin, C. 1859. *The origin of species by means of natural selection.* London: John Murray.

——— 1880. *The power of movement in plants.* London: John Murray.

Delisle, A. F. 1937. "The influence of auxin on secondary branching in two species of Aster." *Am. J. Botany* 24: 159–167.

deWaard, J. and J. W. Roodenburg. 1948. "Premature flower bud initiation in tomato seedlings caused by 2,3,5-triiodobenzoic acid." *Koninkl. Ned. Akad. Wetenschap., Proc., Ser. C.* Amsterdam, 51: 248–251.

Dostál, R. 1908. "Korelační vztahy u klíčních rostlin Papilionaceí." *Rozpr. Česk. Akad. Věd.* II. 17: 1–44.

——— 1909. "Die Korrelationsbeziehung zwischen dem Blatt und seiner Axillarknospe." *Ber. deut. botan. Ges.* 27: 547–554.

——— 1911. "Zur experimentellen Morphogenese bei *Circaea* und einigen anderen Pflanzen." *Flora* 103: 1–53.

——— 1918. "O formativní činnosti reservních orgánů rostlinných." *Rozpr. Česk. Akad. Věd.* II. 26: 1–35.

——— 1925. "O periodě odpočinku u *Circaea intermedia.*" *Acta Soc. Sci. Nat. Morav.* 2: 93–144.

——— 1926. "Über die wachstumsregulierende Wirkung des Laubblattes." *Acta Soc. Sci. Nat. Morav.* 3: 83–209.

——— 1928. "Sur les organes réproducteurs dans *Caulerpa prolifera.*" *Compt. Rend. Acad. Sci.* Paris, 187: 569.

——— 1936. "Die Korrelationswirkung der Speicherorgane und Wuchsstoff." *Ber. deut. botan. Ges.* 54: 418–429.

——— 1941. "Über das Frühtreiben der Fliederzweige, *Syringa vulgaris,* und der Kartoffelknollen, *Solanum tuberosum* durch Verletzung und die hormonale Deutung dafür." *Gartenbauwissenschaft* 16: 195–206.

——— 1941. "Wuchsstoffstudien betreffend die Korrelationen zwischen Wurzel und Spross bei *Pisum sativum.*" *Acta Soc. Sci. Nat. Morav.* 13: 1–31.

——— 1943. "Über die Nekrobiose der Kartoffeldunkelkeime." *Phytopathol. Z.* 14: 484–496.

——— 1944. "Die Anisophyllie der Seitensprosse als Hemmungserscheinung." *Ber. deut. botan. Ges.* 61: 238–249.

——— 1954. "O příčinách dichasiálního sympodia šeříku, *Syringa vulgaris*." *Acta Acad. Sci. Cech. Basis Brunen.* 26: 1–44.

——— 1955. "O korelační morfogenesi na příkladě lnu, *Linum usitatissimum*." *Acta Acad. Sci. Cech. Basis Brunen.* 27: 193–267.

——— 1956. "Značenije korrelativnych vlijanii korněj i listěc v morfogenese rastěnij." *Fiziol. Rast.* 3: 355–367.

——— 1958. "Experiments in plant polarity." *Stud. Plant. Physiol.* Prague, 3: 33–42.

——— and V. Morávek. 1925. "Das Sachssche Phänomen bei Knollen." *Ber. deut. botan. Ges.* 43: 1–10.

Engels, F. 1952. *Dialektika přírody.* Prague: Svoboda.

Errera, L. 1906. "Conflits de préséance et excitations inhibitoires chez les végétaux." *Bull. Soc. Roy. Bot. Belgique* 42: 27.

Fischnich, O. 1939. "Weitere Versuche über die Bedeutung des Wuchsstoffs fur die Adventivspross und Wurzelbildung." *Ber. deut. botan. Ges.* 57: 123–134.

Foster, A. S. 1928. "Salient features of bud-scale morphology." *Biol. Rev. Cambridge Phil. Soc.* 3: 123–164.

——— 1929. "Investigations on the morphology and comparative history of development of foliar organs. I. *Aesculus hippocastanum*." *Am. J. Botany* 16: 441-501.

Galston, A. W. 1947. "The effect of 2,3,5-triiodobenzoic acid on the growth and flowering of Soybeans." *Am. J. Botany* 34: 356–360.

Gäumann, E. 1935. "Über den Stoffwechsel der Buche, *Fagus silvatica*." *Ber. deut. botan. Ges.* 53: 366–377.

Glushchenko, I. L. 1957. "Sovremennoje sostojanie voprosu o vegetativnoj gibridizacii." *Znanije* sér. 8, no. 52, 1–31.

Goebel, K. 1880. "Beiträge zur Morphologie und Physiologie des Blattes." *Botan. Ztg.* 38: 801–815, 817–826, 833–845.

——— 1908. *Einleitung in die experimentelle Morphologie der Pflanzen.* Leipzig, Berlin: B. G. Teubner.

Goethe, J. W. 1790. *Versuch die Metamorphose der Pflanzen zu erklären.* Gotha.

Goodwin, R. H. 1937. "The role of auxin in leaf development in *Solidago* species." *Am. J. Botany* 24: 43–51.

Gunckel, J. E., K. V. Thimann, and R. H. Wetmore. 1949. "Studies of development in long shoots and short shoots of *Ginkgo biloba*

L. IV. Growth habit, shoot expression, and the mechanism of its control." *Am. J. Botany* 36: 309–318.

Haberlandt, G. 1913. "Zur Physiologie der Zellteilung." *Sitz-Ber. preuss. Akad. Wiss.* 318–345.

Haeckel, E. 1860. *Generelle Morphologie der Organismen.* Berlin.

Hagemann, A. 1931. "Untersuchungen an Blattstecklingen." *Gartenbauwissenschaft* 6: 109–155.

Hales, S. 1727. *Vegetable staticks.* London.

Hamner, K. C. 1942. "Hormones and Photoperiodism." *Cold Spring Harbor Symp. Quant. Biol.* 10: 49–59.

——— and J. Bonner. 1938. "Photoperiod in relation to hormones as factors in floral initiation and development." *Botan. Gaz.* 100: 388–431.

——— and E. M. Long. 1939. "Localization of photoperiodic perception in *Helianthus tuberosus.*" *Botan. Gaz.* 101: 81–90.

Hertel, R. and A. C. Leopold. 1963. "Versuche zur Analyse des Auxintransports in der Koleoptile von *Zea mays* L." *Planta* 59: 535–562.

Ille, R. 1937. "Zur Entwicklungsphysiologie der Knospenschuppen bei *Quercus pedunculata.*" *Acta Soc. Sci. Nat. Morav.* 10: 1–19.

Jost, L. 1891. "Über Dickenwachstum und Jahresringbildung." *Botan. Ztg.* 49: 485–495, 501–510, 525–531, 541–547, 557–563, 573–579, 589–596, 605–611, 625–630.

——— 1893. "Über Beziehungen zwischen Blattentwicklung und der Gefässbildung in der Pflanze." *Botan. Ztg.* 51: 89–138.

Klebs, G. 1903. *Willkürliche Entwicklungsänderungen bei Pflanzen.* Jena: G. Fischer.

——— 1913. "Über das Verhältnis der Aussenwelt zur Entwicklung der Pflanzen." *Sitz-Ber. Heidelberg Akad. Wiss., Math.-naturw. Kl.* 55: 1–47.

——— 1914. "Über das Treiben der einheimischen Bäume speziell der Buche." *Abhandl. Heidelberg Akad. Wiss., Math.-naturw. Kl.* 3: 1–116.

Kloz, J. and V. Turková. 1963. "Legumin, vicilin and proteins similar to them in the seeds of some species of the Viciaceae family — a comparative serological study." *Biol. Plant.* 5: 29–40.

Knight, T. A. 1806. "On the direction of the radicle and germen during the vegetation of seeds." *Phil. Trans. Roy. Soc. London,* part I, 96, 99–108.

Kögl, F. and J. Haagen-Smit. 1931. "Über die Chemie des Wuchsstoffs." *Koninkl. Ned. Akad. Wetenschap., Proc., Ser. C.* Amsterdam, 34: 1411–16.

Kořínek, J. 1922. "Sur la sensibilité des corrélations chez les plantes." *Bull. Intern. Acad. Sci.* Prague, 1: 1–6.

Krenke, N. P. 1932. *Wundkompensation, Transplantation und Chimären bei Pflanzen.* Berlin: J. Springer.

——— 1933–1935. *Fenogenetičeskaja izmenčivost* I. Moscow: Biol. Inst. im. K. A. Timirjazeva.

——— 1950. *Teoria cikličeskogo starenija i omoloženija.* Moscow: Selchozgiz.

Kursanov, A. L. 1952. "Dviženije organičeskich veščevstv v rastěnii." *Botan. Zh.* 57: 585–593.

——— 1953. "Vzajmosvjaz fiziologičnych procesov v rastěnii." *Usp. Sovrem. Biol.* 2: 13–25.

——— 1957. "Korněvaja sistema rastěnij kak organ obměna veščevstv." *Izv. Akad. Nauk SSSR* 6.

Libbert, E. 1954. "Zur Frage nach der Natur der korrelativen Hemmung." *Flora* 141: 271–297.

——— 1955. Die Hydrolyse des Korrelationshemmstoffes." *Planta* 40: 256–271.

Lindemuth, H. 1901. "Das Verhalten durch Kopulation verbundener Pflanzenarten." *Ber. deut. botan. Ges.* 19: 515–529.

Loeb, J. 1906. *The dynamics of living matter.* New York. 233 pp.

——— 1924. *Regeneration from a physicochemical viewpoint.* New York: McGraw-Hill.

Lubimenko, V. M. 1910. "Soderžanije chlorofilla v chlorofilnom zerně e energii fotosinteze." *Trudy St. Petersburg Ob-va jestěstvoizpyt.* 41: 97–123.

Lysenko, T. D. 1952. *Stadijnoje razvitije rastěnij.* Moscow: Selchozgiz.

——— 1954. *Agrobiology.* Moscow: Foreign Languages Publishing House. 636 pp.

Mattirollo, L. 1900. "Sulla influenza che la estirpazione dei fiori esercita sui tubercoli radicali delle piante Leguminose." Genoa: *Malpighia* 13.

Maximov, N. A. 1946. "O mechanizmě dějstvija rostovych veščevstv na rastitelnyje klětky. *Bjul. Mosk. ob-va izp. Prirody.* 51: 5–12.

Michurin, I. V. 1948. *Sobrannyje sočinenija.* Moscow.

——— 1950. *Selected works.* English translation. Moscow.

Molotkovsky, G. Ch. 1954. *Zakon poljarnosti rozvitku roslin. Nauk. zap.* Černigovsk. un-tu. 15, IV.

Morgan, T. H. and M. Moszkovski. 1907. *Regeneration*. Leipzig: W. Engelmann.

Moshkov, B. S. 1936. "Rol listěv v fotoperiodičeskoj reakcii rastěnij." *Soc. rastěnijevodstvo* 17: 25–30.

—— 1939. "Transfer of photoperiodic reactions from leaves to growing parts." *Compt. Rend. Acad. Sci. URSS* 24: 485–491.

Mothes, K. 1956. "Die Wurzel der Pflanzen — eine chemische Werkstatt besonderer Art." *Abhandl. deut. Akad. Wiss. Berlin* 5: 24–36.

Müller, F. 1864. *Für Darwin*.

Němec, B. 1900. "Über die Art der Wahrnehmung des Schwerkraftreizes bei den Pflanzen." *Ber. deut. botan. Ges.* 18: 241–245.

——— 1905. *Studien über Regeneration*. Berlin: Gebr. Bornträger.

——— 1908. "Einige Regenerationsversuche an *Taraxacum*-Wurzeln." *Festschr. J. Wiesner*, Vienna.

——— 1911. "Další studie o regeneraci. IV." *Rozpravy Česk. Akad. Věd* 20: 1–20.

——— 1924. "Methoden zum Studium der Regeneration der Pflanzen." *Abderhalden's Handbuch Biol. Arbeitsmeth.* 11: 801–838.

——— 1930. "Bakterielle Wuchsstoffe." *Ber. deut. botan. Ges.* 48: 72–74.

——— 1932. "Über die Gallen von *Heterodera Schachtii* auf der Zuckerrübe." *Stud. Plant. Physiol. Lab.* Prague, 4: 1–22.

——— 1934. "Heterofylie a heterotropie břečtanu, *Hedera helix*." *Rozpravy Česk. Akad. Věd* 44: 1–17.

——— 1934. "Ernährung, Organogene und Regeneration." *Stud. Plant. Physiol. Lab.* Prague, 4: 1–34.

——— 1955. *Dějiny ovocnictví*. Prague: Naklad. Česk. Akad. Věd.

——— 1956. *On the problem of the origin and phylogenetic development of angiosperms*. Prague: Národ Museum.

Noll, F. 1900. "Über den bestimmenden Einfluss der Wurzelkrümmungen auf Entstehung und Anordnung der Seitenwurzeln." *Landwirtsch. Jahrb.* 39: 361.

Pavlov, I. P. 1952. "Polnoje sobranije sočinenij." III. 2: 188–190.

——— 1955. *Sočiněnija*. Leningrad: Gospolitizdat.

Penzig, O. 1921. *Pflanzenteratologie*. II. Berlin: Gebr. Bornträger.

Plett, W. 1911. "Untersuchungen über Regenerationserscheinungen an Internodien." *Dissert.* Hamburg, 1–8.

Podešva, J. 1940. "Über die Wuchsstoffabhängigkeit der Knollen- bezw. Rübenentwicklung bei *Brassica*, *Raphanus* und *Beta*." *Acta Soc. Sci. Nat. Morav.* 12: 1–20.

——— 1947. "Factors preventing the flower-fall in potato, *Solanum tuberosum*. *Acta. Soc. Sci. Nat. Morav.* 18: 1–10.

Prévot, P. C. 1939. "La néoformation des bourgeons chez les végétaux." *Mém. Soc. Roy. Sci. Liège.* IV. 3: 171–340.

Razdorsky, V. F. 1955. *Architektonika rastěnyi.* Moscow: Sov. nauka.

Razumov, V. J. 1931. "O lokalizacii fotoperiodičeskogo razdraženija." *Trudy prikl. genet. i selekc.* 27: 219–279.

——— 1937. "Polučenije klubněj u trudnoklubněobrazujuščich vidov kartofelja putem privivok." *Jarovizacija* 4: 76–88.

Robbins, W. J. 1960. "Further observations on juvenile and adult Hedera." *Am. J. Botany* 47: 485–491.

Sabinin, D. A. 1940. Mineralnoje pitanije rastěnyi. Moscow: *Izv. Akad. Nauk SSSR.*

——— 1957. "O ritmičnosti strojenija i rosta rastěnyi." *Botan. Zh.* 42: 991–1010.

Sachs, J. 1880. "Stoff und Form der Pflanzenorgane." *Arb. Botan. Inst. Wurzburg* II, 452.

——— 1892. *Physiologische Notizen.* Jena.

Sachs, T. and K. V. Thimann, 1966. "Release of lateral buds from apical dominance." *Nature* 201: 939–940.

Schaffner, J. H. 1930. "Sex reversal and the experimental production of neutral tassels in *Zea mays*." *Botan. Gaz.* 90: 279–298.

Schleiden, M. J. 1838. *Beiträge zur Phylogenie.* Berlin.

Schwann, Th. 1838. *Mikroskopische Untersuchungen über die übereinstimmung der Tiere und Pflanzen.* Berlin.

Šebánek, J. 1952. "O stimulaci bramboru a slunečnice kyselinou 2,3,5-trijodbenzoovou." *Sb. Česk. Akad. Zeměděl. Věd. Rostlinná Výroba* 25: 101–116.

——— 1955. " 'Vyskočky' pšenice a kyselina trijodbenzoová." *Česk. Biol.* 4: 630–637.

——— 1956. "O korelativní stimulaci zeleniny směsí kyseliny betaindolyloctové a nikotinové." *Sb. Vysokě Školy Zeměděl. Brno.* 1: 15–21.

——— 1957. "O vlivu trijodbenzoové kyseliny na větvení a kvetení silážní slunečnice za různych podmínek stradijního vývoje." *Sb. Vysokě Školy Zěmědel. Brno.* 3: 441–458.

Sechenov, J. M. 1952. *Izbrannyje sočiněnija. Fiziologija i psichologija.* Moscow, I.

Severcov, A. N. 1911. *Die morphologischen Grundlagen der Evolution.* Jena: G. Fischer.

Skoog, F. and C. O. Miller. 1957. "Chemical regulation of growth and organ formation in plant tissues cultured *in vitro*." *Symp. Soc. Exp. Biol.* no. 11, 118–130. New York and London: Acad. Press.

Sladký, Z. 1957. "Lze evoluci rostlin řešit experimentálně morfologicky?" *Ziva* 5: 162–163.

Slováková, L. 1957. "Průzkum otázky zvýšení odolnosti kukuřice vlivem nízkých teplot." *Kand. diss. Vysoké Školy Zeměděl. Brno.*

Snow, R. 1929. "The young leaf as the inhibiting organ." *New Phytologist* 28: 345–358.

——— 1933. "The nature of cambial stimulus." *New Phytologist* 32: 288–296.

Sorokin, H. P., S. N. Mathur, and K. V. Thimann. 1962. "The effects of auxins and kinetin on xylem differentiation in the pea epicotyl." *Am. J. Botany* 49: 444–454.

Sorokin, H. P. and K. V. Thimann. 1964. "The histological basis for the inhibition of axillary buds in *Pisum sativum*." *Protoplasma* 59: 326–350.

Späth, H. 1912. *Der Johannistrieb*. Berlin: P. Parey.

Stoletov, V. J. 1957. *V nutrividovyje prevraščenija i ich charakter.* Moscow: Sov. nauka.

Thimann, K. V. 1936. "Auxins and the growth of roots." *Am. J. Botany* 23: 561–569.

——— 1939. "Auxins and the inhibition of plant growth." *Biol. Rev. Cambridge Phil. Soc.* 14: 314–337.

——— and F. Skoog. 1933. "Studies on the growth hormones in plants. III. The inhibiting action of the growth substance on bud development." *Proc. Nat. Acad. Sci. U.S.* 19: 714–716.

Timiriazev, K. A. 1922. *Istoričeskij metod v biologii*. I–VI. Moscow.

——— 1943. *Istoričeskij metod v biologii*. Desjat obšče dostupnych lekcij. Moscow.

——— 1951. *Historická metoda v biologii*. Prague: Přírod. vyd.

Troll, W. 1943. *Vergleichende Morphologie der höheren Pflanzen*. Berlin: Gebr. Bornträger, I.

Tumanov, J. J. and G. Z. Garjejev. 1951. "Vlijanije organov plodonošenija na materinskoje rastěnije." *Trudy Inst. fiziol. rast.* im-K. A. Timirjazeva, *Akad. Nauk SSSR* 7: 22–108.

Vaníček, V. 1950. "Antagonismus dusíku a kyseliny 2,3,5-trijodbenzoové při tvorbě růžice u květáku." *Sb. Česk. Akad. Zěmědel. Věd.* 22: 1–7.

Verworn, M. 1901. Allgemeine Physiologie. Jena: G. Fischer.

Virchow, R. 1871. *Die Zellularpathologie*. Berlin: A. Hirschwald.

Vöchting, H. 1878, 1884. *Über Organbildung im Pflanzenreich.* Bonn: M. Cohen.

——— 1892. Über Transplantation am Pflanzenkörper. Tübingen: H. Laup.

——— 1900. "Zur Physiologie der Knollengewächse. Studien über vicariierende Organe am Pflanzenkörper." *Jahrb. wiss. Botan.* 34: 1–148.

——— 1906. "Über Regeneration und Polarität bei höheren Pflanzen." *Botan. Ztg.* 64: 101–147.

——— 1908. *Experimentelle Untersuchungen zur Anatomie und Pathologie des Pflanzenkörpers.* Tübingen: H. Laup. 308 pp.

Vries, H. de (trans. H. Klebahn). 1906. *Arten und Varietäten.* Berlin: Gebr. Bornträger.

Wardlaw, C. W. 1952. *Phylogeny and morphogenesis.* London: Methuen.

Wareing, P. F., C. E. A. Hanney, and J. Digby. 1964. In *Formation of Wood in Forest Trees*, ed. M. H. Zimmermann. New York: Academic Press, 323–344.

Went, F. A. F. C. 1930. "Über wurzelbildende Substanzen bei *Bryophyllum calycinum.*" *Ztschr. f. Botan.* 23: 19–26.

Went, F. W. 1926. "On growth-accelerating substances in the coleoptile of *Avena sativa.*" *Koninkl. Ned. Akad. Wetenschap., Proc. Ser. C.* Amsterdam, 30: 10–19.

——— 1938. "Specific factors other than auxin affecting growth and root formation." *Plant Physiol.* 13: 55–80.

——— 1941. "Polarity of auxin transport in inverted *Tagetes* cuttings." *Bot. Gaz.* 103: 386–392.

Wetmore, R. H. and G. Morel. 1949. "Growth and development of *Adiantum pedatum* L. on nutrient agar." *Am. J. Botany* 36: 805–806.

Wettstein, R. von. 1933. *Handbuch der systematischen Botanik.* Leipzig, Vienna: Deuticke.

Wickson, M. W. and K. V. Thimann. 1958, 1960. "The antagonism of auxin and kinetin in Apical Dominance. I and II." *Physiol. Plant.* 11: 62–74; 13: 539–554.

Wiesner, J. 1905. "Über die korrelative Transpiration mit Hauptrücksicht auf Anisophyllie und Phototrophie." *Sitz-Ber. Akad. Wiss. Wien* 114: 477–496.

Winkler, H. 1907. "Über Pfropfbastarde und pflanzliche Chimären." *Ber. deut. botan. Ges.* 25: 568–576.

——— 1908. "*Solanum tübingense*, ein echter Pfropfbastard zwischen

Tomate und Nachtschatten." *Ber. deut. botan. Ges.* 26: 595–608.

Zalensky, V. R. 1904. "Materialy k količestvennoj anatomii različnych listěv samych těchže rastěnij." *Izv. Kijevsk. politechnič.* in-ta 4: n.1.

Zimmerman, P. W. and A. E. Hitchcock. 1942. "Flowering habit and correlation of organs modified by 2,3,5-triodobenzoic acid." *Contrib. Boyce Thompson Inst.* 12: 491–496.

Index

Note: Numbers in heavy type indicate where Figures may be found.